ENVIRONMENT & POLICY

VOLUME 20

Change and Continuity in Poland's Environmental Policy

by

Magnus Andersson

Institute for Environmental Studies,
Vrije Universiteit, Amsterdam, The Netherlands
& Swedish-Polish Association
for Environmental Protection, Sweden

KLUWER ACADEMIC PUBLISHERS
DORDRECHT / BOSTON / LONDON

A C.I.P. Catalogue record for this book is available from the Library of Congress.

DOI 10.1007/978-94-011-4519-0

Published by Kluwer Academic Publishers,
P.O. Box 17, 3300 AA Dordrecht, The Netherlands.

Printed on acid-free paper

Preface and acknowledgements

This book deals with environmental policy in Poland during the transition from socialism to market-based democracy. The link between environmental issues and economic and political innovation has been evident both before and after this change in political and economic systems. In the 1980s environmental policy was made a test case for political pluralism and market-oriented policy instruments. In the 1990s the harmonization of Polish environmental legislation in conformity with EU directives has appeared as a key task in Poland's strategy for becoming a member of the European Union.

In many countries Poland has the reputation of being the "dirty man of Europe". There are certainly many "hot spots" in Poland but this is not the whole truth about the Polish environment. I have learned that there are huge areas of unspoiled nature in Poland such as the *Bieszczady* mountains in the south, the Baltic coastline in the north, and the so-called Green Lungs in the northeast.

During my visits to Poland I have met many policy makers, researchers and representatives of the private sector and environmental organizations. I am impressed by the skills these people represent. There is enough human capital in Poland to make a contribution towards the transition to a sustainable development in Europe.

I would like to express my gratitude to all those people who in different ways have helped me to conduct research on environmental policy in Poland. Rob Hoppe and Pier Vellinga have provided me with support and valuable comments throughout the project. Matthijs Hisschemöller has spent much time on my research and his support and advice was indispensable for making this ship sail.

Jan Eberg, Joyeeta Gupta, Paul Sabatier, Jürgen Salay, and Harmen Verbruggen have read earlier drafts of the book and provided me with useful comments. I am grateful to Willem Kakebeeke for sharing with me his knowledge and insights about environmental politics in Poland.

I would like to thank *all* colleagues at IVM for creating a friendly and creative working climate. Reyer Gerlagh and Fons Groot had the patience to solve all of my computer problems. Els Hunfeld and Ron Janssen helped me to solve many practical problems inside *and* outside the Institute.

The work I have carried out in Poland would not have been possible without the support of Tomasz Żylicz of Warsaw Ecological Economics Centre (WEEC). The WEEC was my base when I stayed in Poland to collect information and to make interviews. There are many other persons in Poland who I would like to thank for their hospitality during my visits in Poland. To mention a few of them: Piotr Grzegorczyk, Asia Grzywaczewska, Piotr Perszewski, Mariusz Sekta, Janina and Władysław Sobolewski, and the staff of Jumper International and the Institute of the Ecology of Industrial Areas in Katowice.

My thanks to Lars Rydén, Christian Andersson, Fredrik Degerbeck, Magnus Lehman, Paula Lindroos, and Maria Winkler and the rest of the staff of the Baltic University Programme in Uppsala, Sweden, for all sorts of support. I am also grateful for the fruitful cooperation I have had with the Swedish-Polish Association for Environmental Protection.

The Netherlands Scientific Organization (NWO) provided me with the financial assistance for the research. I am very grateful for this.

To my wife, Alicja, and my parents, Marianne and Karl-Erik, I direct my thanks for their continuous support throughout this project.

TABLE OF CONTENTS

List of Figures

List of Tables

List of Abbreviations

ACF	Advocacy Coalition Framework
AWS	Akcja Wyborcza Solidarność (Solidarity's Election Action)
BAT	Best Available Technology
BATNEEC	Best Available Technology Not Entailing Excessive Cost
BOD	Biological Oxygen Demand
BORE	Biuro Obsługi Ruchu Ekologicznego (Service Bureau for the Environmental Movement)
Bq	Becquerel
CBOS	Centrum Badań Opinii Społecznej (Centre for Investigation of Public Opinion)
CFC	Chlorofluorocarbons
CIS	Commonwealth of Independent States
CMEA	Council for Mutual Economic Aid
CUGW	Centralny Urząd Gospodarki Wodnej (Central Water Management Board)
DFES	Debt-for-Environment Swap
EAC	Efficiency Advocacy Coalition
EAERE	European Association of Environmental and Resource Economists
EAP	Environmental Action Programme for Central and Eastern Europe
EBRD	European Bank for Reconstruction and Development
EDF	Environment Defense Fund
EEC	European Economic Community
EIA	Environmental Impact Assessment
EMAS	Environmental Management and Auditing Systems
EMEP	Environmental Monitoring and Evaluation Programme
EPA	Environmental Protection Agency
EPAC	Environmental Protection Advocacy Coalition
EU	European Union
FAO	United Nations Food and Agricultural Organization
FGD	Flue Gas Desulphurization
FEWE	Fundacja Efektywnego Wykorzystania Energii (Foundation for Effective Energy Use)
FYR	Former Yugoslav Republic
GAC	Green Advocacy Coalition
GDP	Gross Domestic Product
GDR	German Democratic Republic
GEF	Global Environmental Facility
GNP	Gross National Product
GUS	Główny Urząd Statystyczny (Main Statistical Office)
HDI	Human Development Indicators
HIID	Harvard Institute for International Development
IPPC	Integrated Pollution Prevention Control
ISD	Institute for Sustainable Development
IUCN	International Union for Conservation of Nature
KERM	Komitet Ekonomiczny Rady Ministrów (Economic Committee of the Council of Ministers)

KKP	Krajowa Komisja Porozumiewawcza (National Coordination Commission)
LOP	Liga Ochrony Przyrody (League for Nature Protection)
LRTAP	Convention on Long-Range Transboundary Air Pollution
MAC	Mainstream Advocacy Coalition
MAGTiOŚ	Ministerstwo Administracji, Gospodarki Terenowej i Ochrony Środowiska (Ministry of Administration, Territorial Management, and Environmental Protection)
MEPNRaF	Ministry of Environmental Protection, Natural Resources, and Forestry
MGTiOŚ	Ministerstwo Gospodarki Terenowej i Ochrony Środowiska (Ministry of Territorial Management and Environmental Protection)
MOŚiZN	Ministerstwo Ochrony Środowiska i Zasobów Naturalnych (Ministry of Environmental Protection and Natural Resources)
MOŚZNiL	Ministerstwo Ochrony Środowiska, Zasobów Naturalnych i Leśnictwo (Ministry of Environmental Protection, Natural Resources, and Forestry)
MW	Megawatt
NCS	Negotiated Compliance Schedules
NFEP	National Foundation for Environmental Protection
NFOŚiGW	Narodowy Fundusz Ochrony Środowiska i Gospodarki Wodnej (National Fund for Environmental Protection and Water Management)
NGO	Non-Governmental Organization
NOT	Naczelna Organizajca Techniczna (Main Technical Organization)
NSW	Nadwiślańska Spółka Węglowa (Nadwiślańska Coal Company)
OBiKŚ	Ośrodek Badań i Kontroli Środowiska (Centre for Measurement and Control of the Environment)
OECD	Organization for Economic Cooperation and Development
OPSZ	Ośrodek Prac Społeczno-Zawodowych (the Social and Union Research Centre of Solidarity)
OPZZ	Ogólnopolskie Poruzumienie Zwiazków Zawodowych (All-Polish Agreement for Trade Unions)
OUG	Okręgowy Urząd Górniczy (Regional Mining Authority)
PAN	Polska Akademia Nauk (Polish Academy of Sciences)
PAP	Polska Agencja Prasowa (Polish News Agency)
PARGWK	Państwowa Agencja Restrukturyzacji Górnictwa Węgla Kamiennego (State Agency for the Reconstruction of the Hard Coal Sector)
PAWK	Państwowa Agencja Węgla Kamiennego (State Agency for Hard Coal)
PHARE	Poland-Hungary Aid for Economic Reconstruction
PIOŚ	Państwowa Inspekcja Ochrony Środowiska (State Inspectorate for Environmental Protection)
PKE	Polski Klub Ekologiczny (Polish Ecological Club)
PLN	Polish New Zloty
PPE	Polska Partia Ekologiczna (Polish Ecological Party)
PPP	Polluter Pays Principle

PPZ	Polska Partia Zielonych (Polish Greens' Party)
PRL	Polska Rzeczpospolita Ludowa (Polish People's Republic)
PROP	Państwowa Rada Ochrony Przyrody (State Council for Nature Protection)
PSE	Polski Ścieci Elektrycznej (Polish Power Grid)
PSL	Polskie Stronnictwo Ludowe (Polish People's Party)
PTTK	Polskie Towarzystwo Turystyczno-Krajobrazowe (Polish Tourist Association)
PZPR	Polska Zjednoczona Partia Robotnicza (Polish United Workers Party)
REC	Regional Environmental Center for Central and Eastern Europe
RECIEL	Review of European Community and International Environmental Law
RP	Rzeczpospolita Polska (Poland's Republic)
SAA	State Atomic Agency
SANEPID	Stacja Sanitarno-Epidemiologiczna (Sanitary Epidemiological Service)
SD	Stronnictwo Demokratyczne (Democratic Party)
SLD	Sojusz Lewicy Demokratycznej (Left Democratic Alliance)
SPM	Svensk-Polska Miljöföreningen (Swedish-Polish Association for Environmental Protection)
ŚRE	Śląski Ruch Ekologiczny (Silesian Ecological Movement)
SSP	Second Sulphur Protocol
TACIS	Technical Assistance for the Commonwealth of Independent States
TDS	Total Dissolved Solids Content
TOE	Tonnes of oil equivalent
TOQE	Tonnes of oil quality equivalent
UCPTE	Union Centrale Pour Transmission Electricité (Central Union for Electricity Transmission)
UN	United Nations
UNCED	United Nations Conference on Environment and Development
UNDP	United Nations Development Programme
UNECE	United Nations Economic Commission for Europe
UNEP	United Nations Environment Programme
UNESCO	United Nations Educational, Scientific and Cultural Organization
UOŚiGW	Urząd Ochrony Środowiska i Gospodarki Wodnej (Bureau of Environmental Protection and Water Management)
UOKŚ	Ustawa o Ochronie i Kształtowaniu Środowiska (Statute on the Protection and Shaping of the Environment)
VOC	Volatile Organic Compounds
WCED	World Commission for Environment and Development
WHO	World Health Organization
WiP	Wolność i Pokój (Freedom and Peace)
WOŚiGW	Wydział Ochrony Środowiska i Gospodarki Wodnej (Department for Environment Protection and Water Management)
WSE	Wielkopolski Seminarium Ekologiczny (Wielkopolska Ecological Seminar)

WUG	Wyższy Urząd Górniczy (Highest Mining Authority)
WWF	World Wide Fund for Nature
ZSL	Zjednoczone Stronnictwo Ludowe (United People's Party)
ZUS	Zakład Ubezpieczeń Społecznych (Social Security Institution)

PART I

INTRODUCTION

1. ENVIRONMENT AND TRANSITION IN EASTERN EUROPE

1.1. Introduction

In the past few decades environmental concern has become an increasingly important economic and political issue. It has mobilized scientific innovation, spawned hundreds of international treaties, stimulated the development of new technologies, and created new regulations and agencies (O'Riordan, 1995). It also played a role in the demise of the Soviet system in eastern Europe. Among the issues the socialist system was unable to address, the deteriorating state of the environment was one of the most striking. Many of the green movements that emerged in the 1980s were connected with the political drive for democracy and independence (Massam and Earl-Goulet, 1997). Environmental issues played the role of political ammunition in the sense that unsatisfied demands for an improved environmental situation fuelled demands for general political and economic changes. In Slovakia, the Slovak Union of Landscape and Nature Protection was virtually the only important current of opposition before 1989 (Jehlicka and Kostelecky, 1992). The political dimension of environmental affairs was evident in cases such as the Hungarian protests against the construction of the Gabcikovo-Nagymaros dam on the Danube river[1] (Fitzmaurice, 1996), Estonian protests against phosphorite mining (Salay, 1991), Lithuanian manifestations against the Ignalina nuclear power station,[2] the attempts to improve the air quality situation in the Bulgarian town Ruse[3] (Baumgartl, 1997), the popular protests in Armenia against the Medzamor nuclear power plant and the Nairit Chemical Factory (Dawson, 1995) and, most importantly, the 1986 nuclear catastrophe in Chernobyl which more than any environmental event in the 1980s undermined Soviet authority in eastern Europe.[4]

---

[1] This joint Hungarian-Czechoslovakian project was launched in 1977. Hungarian protests against the project were led by the independent organization the Danube Circle (*Dunakör*) in the second part of the 1980s (Salay, 1991).

[2] In September 1988 several thousand people formed a human chain around this power plant as a protest against its operation. In October 1988, some 600,000 people (16 per cent of Lithuania's population!) signed a petition against Ignalina (French, 1989).

[3] In 1988 the population in Ruse, a city situated on the Danube, began to protest against the chlorine emissions from industries in the Romanian town Giurgiu on the other side of the river. The issue was picked up by intellectuals in Sofia who subsequently created Ekoglasnost, a nationwide and independent environmental organization, which put the government under considerable pressure. Its activities had the effect of being a catalyst for internal struggles within the communist party, something which contributed to the democratic breakthrough in Bulgaria in November 1989 (Baumgartl, 1997).

[4] Environmental issues also played a role in the political and economic transformation of China during the past few decades. For example, Ross (1988: 196-197) argues that critique against a wasteful water-diversion project in Dazhai contributed to the fall of the chairman of the communist party, Hua Goufeng, an adherent of Mao's line, in the 1970s.

The environmental degradation in eastern Europe has resulted in serious threats to public health, forest die-back, nuclear contamination, heavily polluted water courses, land degradation, and economic losses. The process towards systemic change brought great hopes for improved environmental management in this region.

When western European countries and other OECD countries began seriously to address their environmental problems in the 1970s, these countries were stable and rich market-based democracies. For the eastern European countries in transition[5] beginning to deal with the environmental heritage of socialism, the situation is different and more difficult in several respects: (1) the new environmental policies have to be elaborated at the same time as the rules for making policy are in a state of flux; (2) the governments have meagre financial resources; and (3) the governments have to contend with fledgling political systems, slowly emerging market economies, and unsettled government bureaucracies (World Resources 1992-93, 1992).

The general aim of this book is to increase the understanding of environmental policy-making in countries in transition by taking a long-term perspective on environmental policy in Poland. The rationale for including the pre-transition period in the study is that transformation processes cannot be understood without a solid knowledge of the stage(s) that came before. The book will investigate the driving forces for policy change, both before and after systemic change in 1989, and identify elements of change and continuity. Both of these topics have, by and large, been neglected in the literature on environment and transition.

The first chapter of the book has the following structure. Section 1.2 deals with the environmental situation in eastern Europe from various perspectives. Section 1.3 discusses systemic causes of the environmental conditions in the Soviet bloc. Section 1.4 presents the main pillars of the political and economic transition in eastern Europe. Section 1.5 presents the research problem and Section 1.6 summarizes the chapter.

1.2. The environmental situation in eastern Europe

A plethora of books and articles have discussed the state of the environment in eastern Europe and the reasons for the severity in damage.[6] In the 1980s the environmental degradation in the region became well known, both within and outside the Soviet bloc, because of the spread of *glasnost* and a general upsurge for environmental

---

[5] Countries in transition comprise a group of states that initiated a departure from their socialist systems in and after 1989 (y licz, 1996). These are: the Czech Republic, Hungary, Poland, and Slovakia (sometimes referred to as the four Visegrad countries), Romania, Bulgaria, Albania, Bosnia-Herzegovina, Croatia, FYR Macedonia, Serbia-Montenegro, Slovenia, Estonia, Latvia, Lithuania, Russia, Ukraine, Belarus, and Moldova. To this could be added countries in the Caucasus (Georgia, Azerbaijan, and Armenia) and Central Asia (Kazakhstan, Uzbekistan, Tajikistan, Kyrgyzstan, and Turkmenistan) which all used to belong to the Soviet Union.

[6] See, for example, Alcamo, 1992; Andersson *et al.*, 1993; Cole 1995a, b, c; DeBardeleben, 1985; Feshbach and Friendly Jr, 1992; Fisher, 1992; French, 1991; Hedlund, 1990; Hicks, 1996; IUCN, 1991; Jancar, 1987; Mnatsakian, 1992; Nowicki, 1993; Pavlinek, 1995; Peterson, 1993; Pryde, 1991; Salay, 1991; Sätre-Åhlander, 1993; Turnbull, 1991; Weissenburger, 1995; Welfens, 1993; World Bank, 1992a, b; 1993a, b; 1994; and Ziegler, 1987.

issues.[7] Generally speaking, the environmental record of the former Soviet bloc is more serious than the one of western democratic states. Even though many environmental catastrophes did occur in western democracies after the second world war – for example, a large number of oil catastrophes, the smog build-up in London in 1952, which killed some 4,000 people (McCormick, 1992), and the chemical catastrophe in Seveso in Italy in 1976 (von Moltke, 1995) – it is questionable if they are fully comparable with the ones which occurred east of the iron curtain.

Eastern Europe is an enormous land mass stretching from the Baltic Sea in the west to the Ural Mountains in the east, from the Arctic Ocean in the north to the Black Sea and Caucasus in the south. In this huge area it is possible to find everything from ecological disaster zones to relatively undisturbed nature with a unique flora and fauna. The environmental situation could be divided into three main categories. With respect to industrial pollution the situation is somewhat reminiscent of the industrialized areas in western Europe (such as the Ruhr Valley in Germany) and the USA (such as Pittsburgh) in the 1950s and 1960s. Large amounts of air and water pollution together with hazardous waste are produced by obsolete industries in areas such as Upper Silesia in Poland, northern Bohemia in the Czech Republic, the Donetsk basin in Ukraine, and eastern Ural in the Russian Federation.[8] With respect to radioactive emissions and nuclear contamination the situation in eastern Europe, especially the former Soviet Union, is much more serious than anything experienced outside the region. With respect to biodiversity eastern Europe is, generally speaking, in a better position than western Europe because of the fact that many areas have not been exploited economically and are preserved in their natural state. These are the main faces of the environmental situation in eastern Europe. Below follows a more detailed description of them.

Water, air, waste, and soil

Pollution of water, air, and soil has reached dramatically high levels in many areas in eastern Europe. Pollution of the water resources is one of the most serious problems in the region. The treatment of industrial and municipal waste water is under-developed and existing facilities are poorly maintained. It has been estimated that in most eastern European cities the average efficiency of waste water treatment is less than half the EU norms (RECIEL, 1994). The situation is aggravated by the fact that several countries and regions (for example, Poland) are endowed with scarce water resources.

Out of 273,000 villages in the former Soviet Union only 7,000 had waste water treatment systems at the beginning of the 1990s (Feshbach and Friendly, 1992: 125). In Siberia, a serious problem is the leakage from the extraction and transport of oil.

---

[7] Information about the environmental situation in eastern Europe originating before 1989 is not fully reliable because (1) monitoring was poorly developed; (2) the socialist governments were often disinclined to reveal data about the true extent of environmental decline; and (3) the political opposition was inclined to exaggerate the environmental damage in order to discredit the socialist system.

[8] According to Schreiber (1990) the industrial plants in eastern Europe at the eve of systemic change could be divided into three broad categories: (1) An estimated 10 per cent of total factories falls within the category of “the horror stories”. (2) The second category consists of those factories that do not meet western standards but pose a less immediate threat than the plants in the first category. An estimated 80 per cent of the industries belong to this group. (3) The remaining 10 per cent are, by and large, meeting western environmental standards.

According to one estimation, 16 per cent of western Siberia is affected by this problem (Sätre-Åhlander, 1988: 21).

The Caspian Sea receives, via the Volga river, some 40 per cent of Russia's industrial and municipal waste water, something which has caused a sharp decline in the water quality and the amount of fish species (World Bank, 1992c). Large parts of the Baltic Sea and the more shallow part of the Black Sea lack oxygen due to eutrophication.[9] In such areas no higher life forms can survive. Towns like Vilnius and Kaunas in Lithuania and Riga in Latvia are major polluters of the Baltic Sea (Eckerberg, 1994). Many beaches along the Estonian coastline have been closed because of bacteriological pollution (World Bank, 1993a). The largest water pollution problems related to agriculture appear in Hungary. According to one estimation, one third of Hungary's 3,200 villages experience serious nitrate pollution (Salay, 1991: 59). In the late 1980s, the Czechoslovakian government admitted that 70 per cent of the rivers flowing through the country were heavily polluted (French, 1988). Most of the water in the two main rivers in Poland, Odra and Wisa, are considered to be unfit for any kind of use, including industrial production (Andersson, 1990).

In many areas used by the military considerable environmental damage has been observed. For instance, the activities of the Soviet army in Poland led to contamination of ground waters and land degradation. The total damage in Poland alone has been evaluated at about $2 billion (MEPNRaF, 1995: 27).

Salay (1996: 3) gives the following picture of the air pollution situation in eastern Europe in the 1980s:

> The inefficient use of energy and raw material, the bias towards heavy industry in their national production mix, and their dependence on coal and other fossil fuels all contributed to place the centrally planned economies among the world's most pollution intensive countries (...). In the end of the 1980s, per capita emissions of sulphur dioxide in Poland, Hungary, and Czechoslovakia were two to eight times higher than in western Europe. Their sulphur dioxide emissions per unit of GDP were more than ten times higher than that of western Europe.

The emissions of sulphur dioxide and other acidifying substances have a negative impact on human health, ecosystems, and materials. In Poland and the Czech Republic acidification has caused the destruction of several thousands of hectares of forests on the mountain tops. More than 50 per cent of the Czech Republic's forests may have suffered irreversible damage (Ayres, 1995). In some places in this country the deposition of sulphur reaches 100 kg per hectare and year, to be compared with the critical load for sulphur in the most sensitive forest soils which is a maximum of 3 kg per hectare and year.[10] In the Baltic states the highest emissions of sulphur occur in Narva in northeastern Estonia. The power plants at this place are fired by oilshale containing a high level of sulphur, making Estonia the world's highest per capita emitter of sulphur dioxide in the early 1990s (UNCED/Estonia, 1992: 26). At the end of the 1980s there were about 100 towns in the Soviet Union which had a concentration of air pollution exceeding the Soviet ambient quality standards by more than ten times (Pryde, 1991: 19). One of the towns most subjected to air pollution is Norilsk in northern Siberia, the centre for nickel production. Other industrial centres

---

[9] It should be emphasized that eutrophication is partly a natural phenomenon.

[10] "Critical loads refer to the threshold amounts of pollutants falling to earth (deposited), above which damage will occur" (Nilsson, 1987).

in Russia with serious air pollution problems are Novokuznetsk, Magnitogorsk, and Nizhni Tagil (Sätre-Åhlander, 1988: 9-10).

A common, although poorly documented, problem in eastern Europe is hazardous waste. In 1990 only 12 per cent of the Soviet companies handled their hazardous waste in a safe way (Peterson, 1993). In many areas in eastern Europe waste dumps constitute a direct or potential threat to the drinking-water sources. One example is Tarnowskie Góry in Upper Silesia in Poland where toxic substances from a waste dump near an old chemical factory have polluted the underground water reservoir which is used for drinking water. The clean-up costs may be as high as $2 billion for this site alone.[11]

The opened borders between eastern and western Europe after the demise of socialism have increased the possibilities for "dirty" businesses. At the beginning of the 1990s, Greenpeace International revealed sixty-four trade schemes in which Poland was targeted to receive hazardous waste from western countries (World Resources, 1992-93: 67).

Soil erosion is a major problem in eastern Europe and limits in certain instances the expansion of agriculture. For example, in Romania erosion affects more than four million hectares of agricultural land of which one million hectares in excess of 10-15 tonnes per hectare. This should be compared with the restoration capacity of the soil which is only 4 to 5 tonnes per hectare per year (Zarafescu, 1992). In Poland it has been estimated that some 40 per cent of the land area is threatened by erosion (Chmielewski, 1988: 58).

In the most polluted areas in eastern Europe there is higher morbidity and mortality due to respiratory diseases, cancer, and circulatory diseases among the inhabitants (Bochniarz and Toft, 1995). Life expectancy in Russia dropped sharply between 1987 and 1994: from 65 to 57 years for men and from 75 to 71 for women. According to Murray Feshbach, a US based professor of demography, pollution may play a role in 20 to 30 per cent of Russia's deaths (Nelson, 1996). In Romania four million people are affected by the incidence of pollution with up to 1.5 million people under the permanent impact of pollution (Zarafescu, 1992). In Upper Silesia in Poland the frequencies of respiratory diseases and cancer were 30 to 50 per cent above the average in Poland in the 1980s (French, 1988).

In the early 1990s the Environment Action Programme for Central and Eastern Europe (EAP)[12] identified three main risks to human health: (1) exposure to lead and heavy metal dust from smelters and the use of leaded petrol;[13] (2) exposure to particulates from industrial emissions; and (3) drinking-water pollution (World Bank, 1994).

The environmental situation in eastern Europe is affected by the import of pollution from western Europe. Almost one third of the sulphur deposition in Poland originated from the present EU area in 1991. As far as nitrogen deposition is concerned, the

[11] Personal communication with Albert Tien, 1997.

[12] The Environmental Action Programme was elaborated jointly by the EU, the World Bank, and the OECD and was adopted at a Ministerial Conference in Lucerne, Switzerland, in 1993. It provides policy guidelines on priority setting, institutional development, the choice of policy instruments etc. for countries in transition (Baumgartl, 1997).

[13] The Swedish Ministry of Environment claims that some 400,000 children in eastern Europe have developmental retardments due to lead poisoning (Miljödepartementet, 1997: 67).

western part of Germany had a bigger share than Poland itself (Budnikowski, 1992: 80).

Radioactive pollution

The most well-known environmental catastrophe in eastern Europe is the nuclear disaster in Chernobyl in 1986 which affected virtually the whole European continent. In Belarus, which received 70 per cent of the radioactivity (Pryde, 1991), more than one fourth of the agricultural land has become seriously affected.

Unfortunately, Chernobyl was not an exceptional event with respect to radioactive pollution.[14] It is estimated that the Soviet Union dumped radioactive waste corresponding to at least 2.5 million curie in the Kara Sea and the Japanese Sea. By using the sea as a waste dump the Soviet Union got rid of, among other things, eighteen nuclear reactors from submarines and ice-breakers (Broad, 1993). In addition, at least 11,000 large containers with nuclear waste were dumped in the White Sea (Feshbach and Friendly, 1992).

The Soviet industry producing nuclear weapons was a major source of radioactive pollution. The leading Soviet producer of plutonium for nuclear weapons was Mayak, a nuclear complex situated in the secret town Chelyabinsk-40 in eastern Ural. The total accumulated radiation in the region around Mayak has been pegged at one billion curies – *twenty times* the contamination produced by Chernobyl (Peterson, 1993: 5).

In Semipalatinsk in Kazachstan, where the Soviet Union performed nuclear testings until 1989, up to two million people may have been exposed to radiation from the former testing area (Huber *et al.*, 1992).

Nature

Due to lower population density and a generally lower level of economic activity, the state of the environment in many parts of eastern Europe is arguably less serious than it is in western Europe. Many parts of eastern Europe have not been exploited economically and have therefore been preserved in their natural state. The wetlands, forests, and deltas of eastern Europe tend to have a greater biodiversity than those of western Europe. Examples of well-preserved ecosystems are the Rhodope mountains in southern Bulgaria, the Danube delta in Romania and Ukraine, the Volga delta in Russia, and the biosphere reserve in the Sumava mountains in the Czech Republic (World Wide Fund International, 1993). The 700 kilometre long lake Baikal in Siberia contains 800 plant species and 1,550 animal species found nowhere else in the world. An extraordinary amount of wildlife can be found in the 160 nature reserves set up by the Soviet Union. To mention a few of these areas: Wrangel (polar bear), Almaty (snow leopard), and Lazov (Siberian tiger) (Pryde, 1991). In Poland, the ecologists estimate that near-pristine areas ecosystems cover as much as 8.5 per cent of the territory and that another 19 per cent of the area has biological complexes in good ecological shape.[15] This represents an asset that most OECD countries have lost (ylicz, 1991).

---

[14] A near-meltdown occurred at the nuclear power plant in Greifswald, GDR, in 1975 when a fire cut off power to eleven of the twelve cooling pumps (Stetson, 1990).

[15] About 20-25 per cent of the European stork population is living in Poland (Forowicz, 1997). The Wis a, the largest river in Poland, is unregulated and has a diverse flora and fauna (NFEP, 1997).

The richness of nature in eastern Europe is illustrated by a comparison between the amount of wildlife in Sweden, Estonia, and Latvia, countries which have, roughly speaking, the same geographical position and population density (Table 1.1).

Table 1.1 Wildlife in Sweden, Estonia, and Latvia in 1992

Animal	Sweden	Estonia	Latvia
Wolf	10-20	200	200-400
Bear	800	800	(no estimate)
Lynx	500	900	400
Otter	1,000	1,000	4,000
Black stork	0	200 couples	1,000 couples
White stork	0	1,000 couples	2,000-3,000 couples

Source: Hägerhäll, 1992.

Global impacts

At the time of the demise of socialism in eastern Europe, the region's contribution to environmental problems of global importance was considerable. The eastern European countries accounted for 26 per cent of all carbon dioxide emissions and 17 per cent of chlorofluorocarbons (CFC) (French, 1991). Due to economic decline and other factors, the region's emissions of carbon dioxide have been reduced by 20 per cent since 1988 (Miljödepartementet, 1997: 144).

Economic losses

Environmental damage in Hungary has been estimated at 4-5 per cent of the annual GDP (Reeves, 1995). In the Czech Republic the figure is somewhat higher, 5-7 per cent (French, 1991). The highest cost estimations of Chernobyl reach several hundred billion US dollars. In 1996, Belarus used 3 per cent of GDP to mitigate the consequences of the Chernobyl accident (OECD, 1997). In recent years, Ukraine has devoted 4 per cent of its annual budget to problems related to Chernobyl (Lenssen and Flavin, 1996). In Poland, the annual income loss due to environmental damage was at least 2.5-3 per cent of GDP in the early 1990s (World Bank, 1992b).

1.3. Systemic causes of the environmental decline in eastern Europe

The 1970s saw the birth of the modern concept of environmental protection not only within the OECD but also in eastern Europe. Although the 1972 UN Conference on the Human Environment in Stockholm was boycotted by the socialist bloc (except for Romania) because of an argument over East Germany's voting status at the conference (McCormick, 1992: 97), this event had a tremendous impact on environmental policy development east of the iron curtain. After the Stockholm Conference, several socialist countries promoted the idea of environmental protection in a more systematic way than they had done before, both on the national and international level. National environmental policy strategies were elaborated and an environmental protection committee was set up within the Council for Mutual Economic Aid (CMEA) in 1972 (Aura, 1979a). Another landmark event was the Helsinki Conference on Security and Cooperation in 1975, at which the Soviet Union proposed to reach agreement on three pan-European problems: energy, transport, and environment. The western countries agreed only upon extended cooperation on the last of these themes.[16] The new cooperation was developed within the framework of the UN Economic Commission for Europe (UNECE) and resulted in the 1979 Convention on Long-Range Transboundary Air Pollution (LRTAP) (McCormick, 1992: 183). In the 1980s experts from four socialist countries – China, Hungary, the Soviet Union, and Yugoslavia – participated in the work of the World Commission for Environment and Development (WCED) (Hägerhäll, 1988).

By the mid-1970s all countries of the Soviet bloc had incorporated the goal of environmental protection into their constitutions. The first country to do so was Czechoslovakia in 1960. Later the other countries followed: Romania 1965, GDR 1968, Bulgaria 1971, Hungary 1972, Poland 1976, and the Soviet Union 1977 (Radecki, 1979).

Environmental policy in eastern Europe before 1989 was based on national ambient quality standards which often were more stringent than western ones (Bochniarz and Toft, 1995).[17] Permits for use of natural resources and emission of pollution were issued by environmental agencies on the regional and local level. From the 1970s and onwards, several countries introduced economic instruments such as fees and fines. Also environmental funds, providing grants for various investments for environmental protection and water management, began to operate. For example, within the Soviet Union, the first environment fund was established in Estonia in 1983 (Eckerberg, 1994). Furthermore, environmental protection ministries or agencies were established by several socialist states before systemic change. Politically loyal environmental organizations, mainly concerned with nature protection, were allowed to exist, such as Brontosaurus in Czechoslovakia and the League for Nature Protection (*Liga Ochrony Przyrody* or LOP) in Poland (Salay, 1991).

As shown in Section 1.2, the environmental situation in the Soviet bloc reached a dramatic dimension. Hence, the contrast between policy and performance was stark. Admittedly, environmental degradation is not system specific: it tends to be inherently linked to economic activities of any kind, be it in a capitalist or a socialist country. Economic activities generate several types of pressures on the environment, such as input demands (materials, energy), pollution and waste flows, and spatial intrusion in natural areas (OECD, 1994b: 14). However, economic activities *per se* cannot provide a sufficient explanation. Among authors writing on environmental issues in eastern Europe it is widely assumed that the deplorable state of the environment in this region cannot be understood without reference to the central properties of the socialist

[16] A definition of the dimension of the new cooperation was included in the Final Act of the conference (Hjorth, 1992: 132): "to study, with a view to their solution, those environmental problems which, by their nature, are of a multilateral, bilateral, regional, or sub-regional dimension; as well as to encourage the development of an interdisciplinary approach to environmental problems; to take the necessary measures to bring environmental policies closer together and, where appropriate and possible, to harmonize them".

[17] One example of the strict requirements is that the Romanian water law forbade the pollution of waters in any way (Greenspan Bell, 1992b).

system.[18] The four most important principles of this system were the following: (1) the leading role of the communist party in the society; (2) loyalty to the precepts of Marx and Lenin; (3) state ownership and central planning;[19] and (4) unconditional loyalty to the Soviet Union (Kornai, 1992: 415-416).[20] Below follows a discussion about the way in which these systemic factors may have had a negative impact on the effectiveness of the environmental policies.

The leading role of the communist party severely limited the public's possibilities to influence economic and environmental policies. The lack of freedom for independent environmental organizations and the general public to monitor the behaviour of the polluters spelled environmental disaster. Hence, the leadership of the socialist countries "denied itself the specialist advice and information concerning the extent of ecological decay that non-governmental organizations and environmental movements can provide" (Fagin, 1994: 481).

A central element in the communist party's strategy for keeping control of the political situation was to control the information flow and to portray the socialist system as infallible. Consequently, errors were concealed and perpetuated and failures were not an acceptable mode of learning (Dryzek, 1987).

Enforcement was rendered difficult because of the fact that bureaucracy was not subordinated to the legal system (Kornai, 1992). Party politics was above the law; "law under socialism was a pliable instrument of totalitarian politics and economic policy" (Cole, 1998: 89). In the absence of rule of law every economic sector could simply ignore environmental law.[21]

Loyalty to the precepts of Marx and Engels implied, first of all, that the priority issue for politicians was the class struggle, not environmental protection. Marx and Engels devoted almost no attention to the relation between man and nature. The same holds for the founding fathers of the Soviet system, Lenin and Stalin. Stalin's approach to nature is illustrated by his claim that nature has committed certain "mistakes" that "must be corrected" (Hedlund, 1990: 95).

A number of scholars have emphasized that the socialist pricing mechanism, based on Marx's labour theory of value, led to an underpricing of natural resources.[22] According to Marx, value can only be created by work. Thus, natural resources have

[18] See, for example, Cole (1998), Dryzek (1987), Fagin (1994), Hedlund (1990), Kornai (1992), Peterson (1993), Salay (1991), and Sätre-Åhlander (1993).

[19] The centrally planned economies rested on administrative orders and were managed by a hierarchy with the Politbureau at its head. The government, with its ministries and committees, was responsible for carrying out the decisions of the Politbureau, mainly by issuing binding instructions to subordinate units. Each company was under the authority of a ministry. In Nove's (1980: 62) words, "the Minister and his deputies are super-managers, they are in formal charge of production in the sectors for which they are responsible". Most often, plan-instructions were expressed in physical units. Hence, the role of money, prices, and profits was subordinated (Nove, 1980).

[20] To this could be added the use of coercive institutions like the secret police and the virtual state monopoly of the means of communication (Millard, 1994: 30).

[21] "Lawlessness is the corner-stone of Leninism. Lenin's concept of *Partiinost* (Party supremacy) puts the Party bosses above the law and exposes everyone and everything to the mercy of the Party's whim (...) in the last resort, the Law is the servant of the Party" (Davies, 1986: 40).

[22] See for example Cole (1995c), Hedlund (1990), and Ziegler (1987).

no value until a human hand touches them.[23] This assumption became a typical feature of the socialist price system where the price of, for instance, water and energy, was unrealistically low.[24] Consequently, the socialist economies became highly intensive in energy and resource consumption. In general, the non-market economies in eastern Europe required 2-3 times more input per unit of GNP in the production processes than western economies. This inefficiency was reflected by the emissions of pollutants which typically were 2-3 times higher per unit of GNP than in the OECD (Żylicz, 1992a). Table 1.2 shows that the pollution intensity in eastern Europe was considerably higher than in western Europe in the late 1980s.

The socialist states in eastern Europe pursued aggressive industrialization policies in which large-scale iron and steel mills, power plants, chemical plants etc. were set up at a rapid tempo (French, 1988).[25] For many years the decision-makers in the Soviet bloc regarded smoke from chimneys as proof of progress and prosperity.[26] It is therefore no wonder that measures to reduce air pollution were not undertaken until very recently. The stress on heavy industry led to an under-development of the service sector, that never contributed to more than 30 per cent of the GNP of the economies in the region. This could be compared to the West where the service sector typically provides half of the share of the GNP (World Bank, 1992b). This output-target orientation of the socialist economies was, among other things, a consequence of the material balance approach,[27] which, in turn, was based on Marx theory of value (Arp *et al.*, 1991).[28]

---

[23] It should be emphasized that Marx's theory of value built on ideas developed by the liberal economist Ricardo.

[24] One illustrating example is given by Steenge (1991: 11) who shows that the accounting price charged to producers in primary sectors such as wood felling and mineral extraction was set far too low to enhance resource saving. "These rates were effectively set many years ago and do not at all reflect opportunity costs (i.e. the rate of return on alternative uses). And no rates at all were charged for a number of natural resources, such as coal, oil, and gas reserves, water power, agricultural land, and use of water for irrigation, to name but a few. This means that the financial funds in mining and the basic industries were too scarce to enable enterprises to finance any resource saving systems".

[25] In many cases the technologies were outdated from the outset. For instance, the Lenin Steel Plant in Nowa Huta outside Kraków, which was built in the early 1950s, was based on Soviet technology from the 1930s which in turn was based on American technology from the beginning of this century.

[26] In the West, too, factory smoke was interpreted as a sign of wealth well into the post-World War II period (Sörlin, 1997).

[27] A central element of the planning process was the drafting of material balances. The balance, which is expressed in physical units, is a table identifying sources of supply and uses for various products (Lavigne, 1995: 11).

[28] Arguably, another reason for the rapid industrialization was the wish to increase the military power of the Soviet bloc. In the late 1980s, the Soviet military expenditure absorbed, according to one estimation, up to 17 per cent of GNP while the world average at this time was roughly 6 per cent (French, 1989).

Table 1.2 Economic pressure on the environment in the late 1980s

	CEE-6[a]	EC-12[b]
GDP per capita, $1,000/person	4.8	10.1
Resource intensity of GDP		
Energy, TOE/$1,000	0.77	0.23
Energy, TOQE/$1,000	0.56	0.21
Water, m³/$1,000	153	82
Pollution intensity of GDP		
Industrial solid waste, tonnes/$	1.0	0.4
Waste water, m³/$1,000	83	24
Gases[c], kg/$1,000	51	24
Dust, kg/$1,000	13	1

[a] CEE-6 = Bulgaria, Czechoslovakia, GDR, Hungary, Poland, and Romania.
[b] EC-12 = European Community of the Twelve.
[c] Excluding carbon dioxide.
TOE = tonnes of oil equivalent.
TOQE = tonnes of "oil quality" equivalent: identical with TOE in the case of oil and gas; 2 TOE of coal are assumed to be 1 TOQE; 1 TOE of hydro- and nuclear electricity is assumed to be 3 TOQE.
Source: Żylicz, 1997.

State ownership and central planning. The use of monetary measures in environmental protection was made difficult in an economy that was not monetarized itself and where the enterprises did not have to pay attention to costs. The real value of environmental fees and fines was negligible since "not the money but rather raw material rationing, foreign exchange quota, and other discretional measures determined allocation of resources" (Żylicz, 1992c). Kornai's (1992) analysis of socialist enterprises operating under so-called soft-budget constraints[29] shows that financial performance did not matter because economic losses could always be covered by subsidies from the state. Hence, they could cover the fees and fines with state subsidies or simply ignore them.

The socialist state, as an owner and regulator of the polluting industry, had a regulatory conflict of interest; "the Party/state had a direct financial stake in avoiding environmental law enforcement against the enterprises it owned and controlled" (Cole, 1998: 236).

The Soviet-style hierarchies required long series of interpretations which had a detrimental impact on the implementation process since a certain degree of interpretation is required in implementation. Central planning led to an excessive cognitive load for the planners and provided incentives for the decision-makers to look for large-scale solutions (Dryzek, 1987).

The argument has been advanced that the resource allocation mechanism used in the socialist system, i.e. central planning, was not beneficial for technological innovation. Indeed, industrial technologies were often old-fashioned and inefficient. Also the

[29] In market economies, firms are demand constrained since they are producing for a market and have hard budgetary constraints. When their costs exceed their revenues they go bankrupt. In socialist economies, firms are not constrained by demand as they have soft budgetary constraints. This means that they are always rescued by the authorities if they do not meet their financial targets (Lavigne, 1995: 247).

development of cleaning equipment was delayed in eastern Europe compared to the West (Salay, 1991).

Enforcement was weakened by the fact that the enterprises faced no threat of bankruptcy (Juhasz and Ragno, 1993) and closures of the worst polluters could not be considered because of the prevailing ideology that did not permit unemployment.

Unconditional loyalty to the Soviet Union isolated the eastern European countries in the world economy. Therefore, only to a very small extent did they benefit from the technical developments in the OECD countries. Moreover, while the oil price shocks in the 1970s led to an upsurge for energy saving measures in the West this did not happen in eastern Europe (Nove, 1980).

In conclusion, many systemic causes have been put forward to explain the environmental failure of the Soviet system. It is far from easy to determine which of them are the most relevant ones. This book does not pretend to solve this issue, but it aims at contributing to further exploring them.

1.4. The transition

The systemic change that has taken place in eastern Europe – that is, the shift from central planning to a market economy and from a one-party system to democracy – implies a new framework in which environmental issues are developing. In Poland, systemic change started with the Round-table discussions which were held between the government and the independent labour union Solidarity in the early spring of 1989. The process of systemic change ended with the installation of the first fully democratic government in 1991.

Transition and transformation are two words that have frequently been used in the 1990s in order to describe the political, economic, and social processes in eastern Europe after the demise of the socialist system. It is important to emphasize that the two words are not interchangeable. "Transition" implies change from one state to another. The new state is qualitatively different from the old one (Millard, 1994: 50). Transition and systemic change have the same meaning. The term "transformation" draws attention to the open-ended character of a societal process (Pickel, 1993: 148). Transformation processes occur in all societies, not only in post-socialist ones.

Systemic change in eastern Europe has a political dimension and an economic dimension.

The political dimension

The political dimension of systemic change concerns the introduction of democracy, which, among other things, involves the establishment of democratic institutions and a multiparty system, free elections, freedom of speech, independent courts and mass media, and improved opportunities for public participation in the decision-making processes.

It has been argued that the first phase of the transition offers an opportunity for radical change. It could be seen as, in the words of Leszek Balcerowicz, Poland's Minister of Finance between 1989 and 1991 and again since 1997, a "period of extraordinary politics" (Balcerowicz, 1995) (Figure 1.1):

(...) liberation from foreign domination and domestic political liberalization produce a special state of mass psychology and corresponding political opportunities: the new political structures are fluid and the older political elite is discredited. Both leaders and ordinary citizens feel a stronger-than-normal tendency to think and act in terms of common good (Balcerowicz, 1995: 161).

This period is followed by "the period of normal politics" (Balcerowicz, 1995):

Extraordinary politics (...) quickly gives way to the more mundane politics of contending parties and interest groups (as described, for instance by Nobel laureate James Buchanan and other theorists of the public-choice school). It is in this second period that certain features of political contention which are common in established democracies become much stronger (parties searching for an agenda and an ideological profile, the ensuing politicization of major issues, and so on). These features superimpose themselves on a new democracy (...) (Balcerowicz, 1995: 161).

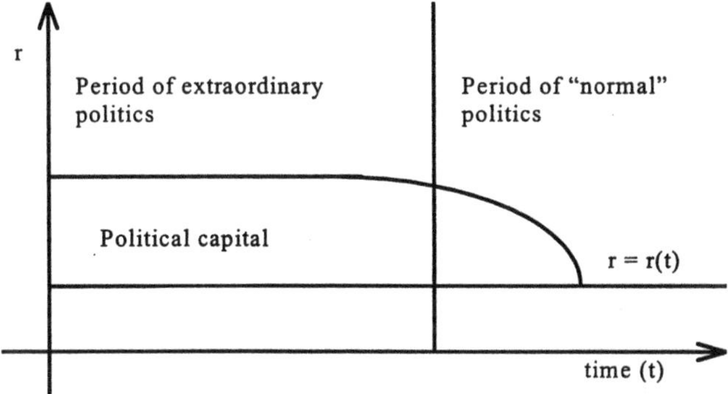

Figure 1.1. A two-stage model of politics. The function r = r(t) expresses the readiness to accept radical policy reform. Based on Balcerowicz (1995).

A consequence of the demise of the socialist system is the necessity to redefine the role of the state. The role of government is related to a whole spectrum of questions concerning the nature of public power, the relation between the public and private sphere, the proper scope of politics, and the appropriate authority of democratic government (Held, 1987). Three questions tend to dominate discussions about the role of government: In what domains of public and private life should the democratic state intervene? What kind of actions should the state undertake? Where are the limits to state action?

The redefinition of the role of government has implications for every sector of society. In the socialist system the state managed virtually all economic activities, while in the new capitalist setting the state plays a more regulatory role. In Poland, there has been a vivid discussion about the role of the government with respect to industrial policy. In the first years of the transition the view of the Ministry of Industry and Trade was that the state should abstain from interfering in the economy and let the invisible hand do the job (Lubbe, 1991). This so-called non-activist approach was later on criticized, by – among others - the ministry itself, for delaying the reconstruction of heavy industry. The dominant assumption of the environmental policy-makers in Poland in the beginning of the 1990s was that the state by all available means should *force* the largest polluters to improve their environmental performance. Presently, there is little support for this approach. These two examples from Poland illustrate that (1) assumptions about the appropriate role of the government may differ depending on the issue area and (2) these assumptions may change over time.

The economic dimension

The economic dimension of systemic change is the transition from state ownership and central planning to capitalism and a market economy. The economics of transition have emerged as an increasingly important theme in the economic literature in the 1990s where issues such as the legacy of the socialist system and the content, speed, and phasing of reform (usually from a macroeconomic perspective) are highlighted.[30] While no blueprints for the design and implementation of the transition existed in 1989 (Yarnal, 1995), today there is a non-trivial degree of agreement in the literature that a typical reform programme should contain at least the following elements:

1. macroeconomic stabilization and control;
2. price and market reform; and
3. private sector development, privatization, and enterprise restructuring.

Macroeconomic stabilization and control aim at curbing inflation, eliminating the so-called monetary overhang, and improving the external trade balance. Important means for achieving these goals are fiscal tightening, tight credit policies, wage controls, elimination or sharp reduction of government subsidies (for example, on food, energy, housing, and transportation), and currency devaluation. *Price and market reform* involves abolishment of central planning, liberalization of retail and wholesale prices, international trade liberalization, reform of financial markets, the establishment of commercial banks, and demonopolization of systems for transport, distribution, and trade. *Private sector development, privatization, and enterprise restructuring* are promoted by establishing private property rights and legalization of private ownership. Privatization of state-owned enterprises could be done on a case-by-case basis, via voucher privatization or the establishment of investment funds managing a group of enterprises. An important role in the privatization process is played by foreign capital that helps to transfer new technologies and management skills (Pavlinek, 1995).

In addition, it should be emphasized that an important task for the state in the first years of the transition period is the establishment of a social safety net in order to, among other things, provide economic compensation to the unemployed and other vulnerable groups in the society (*ibid.*).

Among economists, there has been a debate about the proper speed and radicality of the reform programme. According to the advocates of so-called shock therapy, the best solution is a quick and radical jump from the old system to the new one.[31] As Blanchard *et al.* (1991: x) put it, countries in transition, "have little choice but to move on all fronts at once – or not move at all". Leszek Balcerowicz, the chief designer of the Polish "Big Bang" in 1990, brings forward a psychological argument

---

[30] See, for example, Aghevli *et al.*, 1992; Balcerowicz, 1995; Blanchard *et al.*, 1991; Blanchard *et al.*, 1994; Campbell, 1991; Clague and Rausser, 1993; Claudon and Gutner, 1992; Corbo *et al.*, 1991; Ellman *et al.*, 1993; Gelb and Gray, 1991; Keren and Ofer, 1993; Klaus, 1992; Köves, 1992; Richter, 1992; and Sachs, 1994.

[31] See, for instance, Sachs, 1994 and Balcerowicz, 1995.

for defending shock therapy. In his view, "people are more likely to adapt internally to external changes if they are radical and, therefore, perceived as irreversible, than if they are small and thus regarded as easy to reverse" (1995: 13). Furthermore, the reform programme should be launched in the "period of extraordinary politics" in order to gain support (see Figure 1.1).

Balcerowicz dismisses the point that shock therapy can bring about political turmoil by claiming that there is no simple link between types of economic reform programme and political stability (*ibid.*: 226).

The adherents of the gradualistic approach argue that over-hasty radical reform is doing more harm than good in that it could, for instance, provoke mass popular protests on such a scale that the whole reform process may be in danger. Criticism has been voiced concerning the fact that it is the liberal Anglo-American theories of capitalism and the free market which have been promulgated in eastern Europe, not the ideas of the more interventionist and consensus-seeking Japanese model, nor the pragmatic German social market economy (Keegan, 1993).[32] Chang and Rowthorn (1995: 46) claim that the neo-liberal approaches which have dominated the reform agendas in Poland and other parts of eastern Europe since 1989 have undermined already weakened state structures. In their view, this has had detrimental effects on economic development. The small-scale, low-tech, and non-tradeable areas have been revived but few structural changes, for instance, within heavy industry, have occurred. By and large, eastern Europe therefore still lacks internationally competitive industries.

The success of the transition in eastern Europe depends not only on the policies inside this region but also on western responses. Indeed, "it will take actions on both sides for eastern Europe to return to Europe" (Sachs, 1994: 6). The West's financial support involves both grants and loans from government-to-government as well as loans from multilateral institutions such as the World Bank, the International Monetary Fund (IMF), and the European Bank for Reconstruction and Development (EBRD).[33] Financial and technical assistance is crucial for the success of the reforms, especially in the initial phase. Western countries are also important for eastern Europe as trade partners (*ibid.*: 102).

The differences between the countries that once belonged to the Soviet bloc become increasingly pronounced as the social, economic, and political transformations proceed. These processes take place in concrete places influenced by their history, geography, and culture (Pavlinek, 1995: 4). There is no universal path from socialism to capitalism but rather a number of divergent national and regional pathways. Thus far, the departure from the Soviet system to market-based democracy has clearly proceeded faster in some parts of the region than in others. Countries such as Poland, Hungary, and the Czech Republic are well-established democracies with a relatively

[32] According to Keegan, the fall of the Berlin wall coincided with a period in the West when non-market solutions were increasingly being rejected: "when dissatisfaction with the performance of the market economies became widespread in the 1970s, an opportunity for the New Right and extreme free marketers to gain ascendancy arose. Their ideas, or prejudices, not only led to the wild embrace of 'deregulation' in the 1980s, but also had a pernicious influence on attitudes towards third world aid and shaped the 'big bang', radical economic and social chaos in eastern Europe and the former Soviet Union" (Keegan, 1993: ix).

[33] The cumulated financial flows (grants and non-aid related financial flows such as borrowing on capital markets, foreign direct investments and portfolio investments) from the West to eastern Europe between 1990 and 1993 has been estimated at $167 billion (Lavigne, 1995: 236).

stable economic growth and increasingly close ties with the European Union. At the same time some of the former Soviet Republics in Central Asia and Caucasus have been classified as developing countries by the OECD (Laurson *et al.*, 1995). In addition, in virtuallly all countries of the former Soviet bloc there are disparities among regions and among the cities and the countryside.

Finally, a few remarks about the end-point of the transition. According to some analysts, the period of overcoming socialism will last as long as socialism itself prevailed (Tkaczyński, 1997). They claim that it is difficult to consider the transition process as completed until the cultural and behavioural legacy of the old system is eliminated. It is still an open question what policies will be pursued in the post-transitional economies of eastern Europe. The experience of industrialized capitalist countries does not give a clear-cut answer of what is the proper or "normal" balance between the state and the market. As Kaminski (1994: 3) has noted, "the institutional diversity of modern market economies strikes one as overwhelming". For example, the newly industrialized countries in eastern Asia have in general relied on state interventionism. The governments in this region have directly affected the allocation of resources in the economy through various industrial, trade, and credit allocation policies (Haggard and Webb, 1993: 250). Countries such as the USA and Great Britain have tried to restrict the government's role in the economy. Even within the European Union it is possible to discern different approaches to the organization of economic life among the member countries. Hence, all that can be said is that a whole array of end-points of the transition process is possible (Kaminski, 1994: 3) and that the outcome of the transition in eastern Europe is likely to be a spectrum of different approaches.

1.5. Research problem and focus

This chapter has shown that state socialism in eastern Europe apparently failed to address environmental problems. The question as to what caused environmental degradation can be investigated by comparing environmental policy before and after systemic change. This study will especially look into the degree of change. Policy change can be divided into three categories. First-order change is incremental in nature; change proceeds in small steps. The crucial determinants of policy are previously made decisions. Second-order change involves greater change, for example, a shift in the use of policy instruments, without challenging the existing policy paradigm. Third-order change implies radical change and involves a paradigmatic shift (Dudley and Richardson, 1996: 65). It should be emphasized that change and continuity are not mutually exclusive concepts. On the contrary, change may co-exist with stability (*ibid.*: 67). Another phenomenon is that long periods of continuous incremental change in the end can produce radical change.

The general research problem dealt with in this book could be stated as follows:

To what extent has systemic change in Poland affected environmental policy in the 1990s?

The choice of Poland as the country being in focus of this study has several explanations. First, Poland has, in comparison with its neighbours in eastern Europe, a relatively long tradition of environmental policy with respect to, for example, policy instruments and institutional arrangements. Environmental protection emerged as an important issue on the political agenda already in the 1970s and 1980s. Thus, in Poland it *is* possible to systematically compare environmental policies *before* and *after* systemic change. In addition, the fact that environmental policy was an established policy area in Poland in the socialist period provides an opportunity to precisely define the implications of systemic change from an environmental perspective. In other words: evidence for a relation between systemic change and changes in environmental policy performance should be easiest to find in a country where there was a manifested will to address environmental problems before systemic change. Second, the environmental policy-making process in the 1980s had a number of features that gave Poland a unique position in the Soviet bloc: Poland allowed the existence of one independent environmental organization, the Polish Ecological Club (*Polski Klub Ekologiczny* or PKE); the repression against environmental organizations was milder than the repression against those organizations that formally belonged to the political opposition; there was a certain degree of access to environmental information; and certain interaction between various actors was possible and made the environmental policy process more dynamic than in other countries in the region. Thirdly, Poland is considered to belong to those countries in post-socialist Europe that have the most advanced management system for environmental protection. Hence, the example set by Poland with respect to environmental policy may have broader implications for other countries in transition. Fourth, Poland is one of the largest countries in eastern Europe, with more than 38 million inhabitants, and as such its impact on the environmental situation in Europe is considerable. Fifth, Poland is an associated member of the European Union and is expected to become a full member within five to ten years. As an EU member Poland may soon become an important actor in European environmental policy. Lastly, the author of this book, a Swede by nationality, is familiar with the Polish language and has been involved in various forms of environmental cooperation between Sweden and Poland since 1986.

1.6. Summary

The environment was an important "banner" in the social and political transformation processes in eastern Europe in the latter part of the 1980s that eventually led to the demise of the socialist system between 1989 and 1991. The chapter showed that the environmental performance of the Soviet bloc was worse than that of western European countries. The causes for this are related to the economic and political foundations of the socialist system. There is a variety of systemic factors that can be taken into account.

The first steps in the political and economic transition involve the dismantling of the command economy and establishing free elections and a multi-party system. The precise end-point of the transition is difficult to define and varies between the post-socialist countries. Both political and economic dimensions of the transition are likely to have an impact on environmental policy and performance.

The general research problem of this study is: to what extent has systemic change in Poland affected environmental policy in the 1990s?

Poland has been selected for several reasons. Especially important is that environmental policy in Poland was an established policy area before systemic change, which means that it is possible to systematically compare developments in the 1980s and 1990s.

2. A POLICY SCIENCE PERSPECTIVE

2.1. Introduction

This research on change and continuity in Polish environmental policy is made from a policy science perspective. For the purpose of this study, various insights from this discipline will contribute to an understanding of how and why policies have or have not changed.

Most theories, insights, and models on policy-making and policy processes have been developed from the perspectives of policy analysts working in western democracies. From the 1960s and onwards, a continuous interaction between European and American scholars has resulted in comparative models of policy-making, related to either different political systems and national styles of policy-making (political culture) or to the construction of different types of policy problems addressed by the same political system (Andersson and Hisschemöller, 1997; Hisschemöller, 1993). Although agreement has not been reached on one multiple-perspectives model of the policy process, recent decades have shown a convergence of approaches. One model, which has raised increasing attention among policy scientists, is the advocacy coalition framework (ACF) developed by the American political scientist Paul Sabatier (Sabatier and Jenkins-Smith, 1993). This model deals with the entire policy process and claims to combine different research perspectives, such as the agenda-building approach, implementation, and the analysis of so-called policy belief systems. The main focus of the model is on policy change and policy-oriented learning by coalitions that compete for realizing their objectives within the realm of a so-called policy subsystem such as environmental policy-making. This makes the ACF suitable for the purpose of this book. I will use the ACF to analyse Polish environmental policy in the era of transition to a market-based economy and political democracy. As the ACF is based on western insights of policy-making, environmental policy in Poland may serve as an interesting case to clarify whether the application of this model can be extended to "non-western" cases. Before presenting this model in some detail it is useful to dwell on some of the key concepts used in analysis of public policy which also appear in the ACF. Section 2.2 discusses the main elements of the policy-making process. Section 2.3 is devoted to policy instruments while Section 2.4 presents the concept of belief systems. Section 2.5 addresses the impact of external factors and stable parameters on environmental policy-making. Section 2.6 is devoted to the ACF. It presents the framework and discusses the implications of applying it to a non-democratic setting. Section 2.7 presents the research questions and the structure of the book. Section 2.8 deals with the research design and approach. The last section briefly summarizes the chapter.

2.2. The policy process

It is useful to begin this section on the policy-making process with a clarification: policy should not be mixed with politics. According to van de Graaf and Hoppe (1989: 2) politics should be seen as the struggle over policy. They regard policy as a "politically endorsed plan that can be used to achieve certain goals; a method for preventing or combating problems".[34]

Stages in the policy process

A considerable body of literature on the policy-making process divides this process into a number of distinctive stages. In sum, "[o]ne studies first how policy problems arise and appear on the agenda of government decision making, then how people formulate issues for action, next how legislative action follows, how administrators subsequently implement the policy, and finally at the end of the process, how policy is evaluated" (Lindblom, 1980: 3).

The starting-point for the policy process is the appearance of a problem.[35] A problem can be seen as a situation where there is a gap between a normative standard and a perception of an existing or expected situation. From this it follows that a policy problem, like air pollution, is not an objective condition. Facts, conditions, and situations may be interpreted differently by various people, which means that the same information can result in conflicting perceptions (Hoppe and Peterse, 1993). Indeed, "policy problems are in the eye of the beholder" (Dunn, 1981: 97). In sum, a problem is not a given fact but a socio-political construct (Hoppe and Peterse, 1993).

What is the problem about? What values are at stake? What facts are the relevant ones? These are questions that arise when a problem is structured. Problem structuring has important implications for the policy-making process because "[t]he name for a problem (...) creates beliefs about what conditions public policy can change and which it cannot touch" (Edelman, 1977: 25-26) and "the way people classify a problem determines the way they will attempt to resolve it" (Dunn, 1981: 98). From this follows that the description of problems, explicitly or implicitly, inherently refers to solutions (Hisschemöller, 1994) and that "solutions" are a kind of social construction.

A common feature for dictatorships is the denial of the existence of problems which are not structured. An unstructured problem is characterized by disagreement about both knowledge and values (Hisschemöller and Hoppe, 1996). An elucidating example is the official approach to the relationship between environmental

---

[34] Dunn (1981: 61) sees public policy as "long series of more or less related choices, including decisions not to act, made by governmental bodies and officials." In Kleijn's (1968: 7) words, policy is "acting deliberately and systematically, using the appropriate means, with a clearly defined political objective in mind which is worked towards step by step". Lundqvist (1980: 24) defines the term as a "problem-solving activity involving goal-conscious actors". The following has also been said about the characteristics of policy (Hogwood and Gunn, 1984: 19-23):

- Policy and decision are not identical concepts. A policy usually consists of a series of decisions.
- The making and implementation of policy occurs within the framework of an administration.
- Policy involves both behaviour *and* intentions.

Scholars dealing with public policy generally share the opinion that policy includes action and inaction, decisions and non-decisions. However, few would argue that *everything* a government does not do is a policy. Heidenheimer *et al.* (1990: 5-6) clarify: "Government inaction, or non-decision, becomes a policy when it is pursued over time in a fairly consistent way against pressures to the contrary (...) what matters for purposes of identifying 'policy in repose' is that the issue be perceived by at least some major participants as being on the political agenda."

[35] Problems can be distinguished from "phenomena" in that the former refer to solutions, the latter not (Hisschemöller and Hoppe, 1996). Thus, sulphur emissions from a volcano cannot be considered to be an example of an environmental problem.

degradation and health problems in socialist Poland, especially before the 1980s. Newspapers and environmental journals could not publish articles about this issue. Everything that indicated a connection between environmental degradation and the health status of the population was immediately eliminated by the censorship.[36]

Policy adoption. It is important to emphasize that this stage includes everything from policy documents, like White Papers that are background reports or "papertigers", to strictly binding laws. It is noticeable that there is a much wider range of policy documents of varying degrees of legality in parliamentary systems in western Europe than in the USA.

Implementation. In the implementation stage attempts are made to realize policy. According to *Webster's Dictionary* to implement means "to carry out: to accomplish, fulfil; to give practical effect to and ensure of actual fulfilment by concrete measures, to provide instruments or means of practical expression" (Lane, 1995: 97). In order to translate words into deeds it is necessary to have access to financial resources, personnel, organizational structure etc. However, the activities undertaken in the implementation phase need not lead to the fulfilment of the policy objectives. As has been shown by ample literature on implementation, discrepancies between promise and performance frequently occur (see Pressman and Wildawsky, 1973 and Mazmanian and Sabatier, 1989).

Evaluation. In the evaluation phase, the result of a public programme is assessed with respect to the intended and unintended effects. All sorts of activities undertaken during the policy-making process are evaluated. Mistakes are identified and explained and lessons for future policy-making are drawn (Premfors, 1989: 132-134). However, often policy is not sufficiently evaluated or not evaluated at all.

The agenda-building perspective

A stages model that will be used in this study is the agenda-building perspective. *Agenda-building* has been described as "the process by which demands of various groups in the population are translated into items vying for the serious attention of public officials (...)" (Cobb *et al.*, 1976: 126). The model distinguishes between two agendas: the public agenda, consisting of issues which have achieved a high level of public interest, and the formal agenda, consisting of items that the decision-makers have formally accepted for serious attention. According to Cobb and Elder (1983: 86), there are three prerequisites for an issue to enter the public agenda: widespread attention or at least awareness; a shared concern of a sizeable portion of the public that some type of action is required; and a shared perception that the matter is an appropriate concern for the government.

Changing scientific knowledge, large-scale catastrophes, public pressure, and

[36] A governmental decree from 1970 declared that: "All information pertaining to any immediate threat to life and health of people caused by industry and by chemicals used in agriculture must be eliminated from all reports on environmental protection and on the threat to the natural environment in Poland. This ban applies to specific cases of the pollution of the atmosphere, water, earth, and food dangerous to life and health" (quoted in Cole, 1998: 107).

Krystyna Forowicz, a journalist at the newspaper *Rzeczpospolita*, recalls that "it was not possible to write about diseases, that the children in Silesia were ill because of lead pollution. When I wrote about these children I was told that it was not true. The editor said to me: Krystyna, do you want to frighten people? Where else can they live? Look for good examples instead, for instance, some industrial plants which have installed filters" (personal communication, 1995).

transitory political conditions are examples of factors that can help an issue to enter the political agenda. Non-governmental organizations, media, and professional expertise play an important role as agenda setters. Referring to the last of the three groups, Weale (1992: 60) writes that "some items of policy will not come on to the policy agenda unless it is placed there by experts: we would not even know there was a problem of global climate change without the professional work of climatologists and other scientists".

Thus, it is clear that an issue may be brought to agenda in different ways. Cobb *et al.* (1976: 126-128) distinguish between three agenda-building models: (1) *The outside initiative model,* which is the process in which a non-governmental organization brings up an issue to reach the public agenda and finally the formal agenda. (2) *The mobilization model,* which describes a situation where an issue is initiated inside the government (thus automatically reaches the formal agenda) but where successful implementation requires that the issue also be placed on the public agenda. (3) *The inside initiative model,* which considers an issue that arises and remains within the government.

In his famous article "Up and Down with Ecology - the Issue-Attention Cycle" Downs (1972) made the observation that the intensity of the attention that is paid to an issue on the agenda tends to be reduced as time goes by. The alarmed discovery and euphoric enthusiam of the first phases of the attention cycle tend to be followed by phases characterized by a more pragmatic and less emotional approach to the issues appearing on the political agenda.

A major objection raised to the agenda-building perspective and to the stages model in general is that implementation cannot be seen as a mere instrumental execution of earlier agreed policy. It is argued that the shaping of a policy continues throughout the implementation phase (Bachratz and Baratz, 1970: 61); that the "real decisions" are sometimes taken when policy is realized, not when it is adopted; and that policy-making occurs as bureaucrats attempt to implement vague legislation (Sabatier and Jenkins-Smith, 1999). Kingdon (1984) has made a case for the view that problems, policies, and politics are three independent streams which have their own dynamics and flow according to independent processes. Policy change is most likely to occur when the three streams are coupled. This tends to be the work of a policy entrepreneur who benefits from a short-run opportunity (a "policy window") to highlight a particular problem or solution. Another major finding in Kingdon's work is that policy alternatives tend to be elaborated before the agenda is set. In line with Kingdon, the "garbage can model" (Cohen *et al.*, 1972) sees the decision-making process as an *ad hoc* mixture of problems and solutions. The model is based on the assumptions that the value function is ambiguous, knowledge about the choice situation is uncertain, and decision rules are complex and symbolic. In addition, the stages model has been criticized for not being a causal model, for neglecting the fact that evaluations of existing programmes often affect agenda-setting, and for having a top-down bias which implies that so-called street-level bureaucrats and other actors are excluded from the analysis (Sabatier and Jenkins-Smith, 1993).

It has however been argued that agenda-building focuses on phenomena such as *power relations, non-decisions, and patterns of inclusions and exclusion of issues and actors* rather than on rational management of the policy process (see, for example, Cobb and Elder, 1983 and Hisschemöller, 1993). It is important to emphasize that these concepts may be applied for analysing *all stages* of the policy process. The issue of power was earlier addressed by Schattschneider. In Schattschneider's (1960: 71) view, "all forms of political organization have a bias in favour of some kinds of conflict and the suppression of others because organization is the mobilization of bias. Some issues are organized into politics while others are organized out". The same holds true for the policy actors: some actors are included into the policy process while others are excluded. The literature on power emphasizes that non-decision making frequently occurs in the policy-making process, especially in the agenda-setting phase, but also in the policy adoption and implementation phases. A non-decision, as defined by Bachratz and Baratz (1970: 44), is a "decision that results in suppression or thwarting of a latent or manifest challenge to the values of interests of the decision-maker". An obvious example of a non-decision is the Polish government's approach to industrial policy in the first years of the transition. The Minister of Industry in 1989-91, Tadeusz Syryjczyk, officially made clear that "no industrial policy is also a kind of industrial policy" (Górka, 1994: 13).

Interactions

The interaction between the actors in the policy-making processes can take different forms. The *totalitarian model* relates to political systems in which one group or political party completely dominates the policy process (Ziegler, 1987). This model is associated with dictatorships. The *pluralist model* relates to a politics-of-interests based policy process, where groups in society try to obtain their specific goals and objectives by lobbying and coalitions formation with like-minded interests. Assumptions that underlie this model are that groups and group members who take part in influencing policies know what they want, i.e. what is in their interest, that the competing interests are equally legitimate, that no group of interest is powerful enough to rule on its own, and that the policy preferences of the different groups allow for a trade-off in negotiations with others. Since coalitions shift over time with respect to different issues, it is assumed that in the long run individuals will be compensated for a loss in one area by a benefit in another. Interests are usually represented by elites who do the actual lobbying and bargaining. Together, the elites form a quite stable and homogeneous group and policy change is supposed to be incremental (Braybrooke and Lindblom, 1963). This model of policy-making comes closest to the metaphor of the political market. The US air protection policy could be seen as a typical example of pluralist policy-making in that four different groups – the governmental administration, the environmental movement, industry, and the political elite – are actively influencing the policy process (Bressers and Huiteman, 1996: 173).

Ziegler (1987), who analysed environmental policy-making in the Soviet Union in the 1980s, has pointed out that environmental policy in the Soviet Union cannot be typified as completely totalitarian. In his view, the policy process had *corporatist* features. Where pluralist theories on policy-making start from the assumption that policy actors and institutions compete because they have different interests, the corporatist theory only counts for a very limited degree of interest representation (Schmitter, 1979). Corporatism assumes that the state is an organic whole and that the different compartments (industry, environment, etc.) have the major function to serve the well-being of the state. Corporatism also assumes a hierarchical order between the different compartments.

Corporatism can be divided into societal corporatism and state corporatism. The former refers to a process in which "societal groups are autonomous entities capable of penetrating the state apparatus for their particular individual purposes" (Ziegler, 1987: 46). In societal corporatism emphasis is laid on the inclusion of interest organizations in the policy-making process in order to maximize governmental responsiveness to society (*ibid.*: 75). State corporatism "presumes active state involvement and manipulation of society to achieve a nonpluralist order" (*ibid.*: 46). The actors in the policy process are dependent on the state and penetrated by it. In the state corporatist model it is the state, not the group, that plays the central role in the agenda-building process (*ibid.*: 47-48). Ziegler concludes that it is the state corporatist model that corresponds with the Soviet environmental policy-making style.

A fourth model is the *participatory model*, characterized by self-rule and direct democracy (Held, 1987).

One of the aims of this study is to investigate to what extent totalitarianism, pluralism, societal corporatism, state corporatism, and direct democracy characterized the environmental policy process in Poland before 1989 and whether these forms of interaction are discernible after systemic change.

2.3. Policy instruments

Policy instruments are tools used by policy-makers in their attempts to alter societal processes in such a way that they become compatible with the policy-makers' objectives (OECD, 1994a: 14). In Vedung's (1995: 1) definition, policy instruments are "the set of techniques by which governmental authorities wield their power in attempting to ensure support and effect or prevent social change".[37] Public policy instruments are generally divided into three classes: regulations, economic means, and information/moral suasion.[38]

Regulation (also called command-and-control instruments) comprises a range of direct regulations such as standards, bans, permits, zoning use restrictions, etc. In the OECD's (1994a: 15) definition, direct regulations are "institutional measures aimed at directly influencing the environmental performance of polluters by regulating processes or products used, by abandoning or limiting the discharge of certain pollutants, and/or by restricting activities to certain, times, areas, etc". Vedung (1995: 9) puts it slightly differently: "Regulations are measures undertaken by governmental units to influence people by means of formulated rules and directives which mandate receivers to act in accordance with what is ordered in these rules and directives". Within countries belonging to the OECD, regulation has traditionally been the most commonly used policy instrument in environmental protection.

The second approach is the application of economic instruments to create environmentally appropriate behaviour. The main economic instruments could be categorized as follows: charges and taxes[39] (effluent charges, product charges, and tax

---

[37] According to Klok (1991), an instrument is anything a government has decided to use in order to promote objectives.

[38] Winsemius (1989: 77) names these instruments "the whip, the carrot, and the tambourine and the song". When the whip (direct regulation) is used the desired behaviour is coerced via legislation and rule-making. The carrot (indirect regulation) provides financial incentives for influencing choices. The tambourine and the song refer to "careful interaction with the target groups so much sympathy for the environmental policy goals is created that the target groups assume responsibility for their own actions and make the jump through the hoop without either the whip or the carrot".

[39] The difference between charges and taxes is that the former are rendered in return in the form of a good or a tax while the latter go the general budget without any direct return.

differentiation), subsidies, deposit-refund systems, market creation (emissions trading, liability), and financial enforcement incentives (non-compliance fines and performance bonds) (OECD, 1994a). Economic policy instruments involve either the handing out or the taking away of material resources. In other words, economic instruments make it cheaper or more expensive to pursue certain actions (Vedung, 1995: 10).

Four central concepts in environmental policy are effectiveness, efficiency, cost-effectiveness, and equity. *Effectiveness* concerns the extent to which a measure, such as an investment, succeeds in reducing environmental impacts in relation to policy targets set. *Efficiency* has to do with the extent to which the costs of a policy are justified in terms of its effects and if it maximizes the effects minus the costs (Semeniene and Żylicz, 1997). A *cost-effective* policy seeks the least costly method of attaining a specific environmental quality goal (Downing, 1984: 33). *Equity* relates to the balance between costs and benefits across the parties concerned (Semeniene and Żylicz, 1997). Hence, it has to do with burden-sharing and fairness. It is difficult (but not impossible) to design policies that combine the notions of effectiveness, efficiency, and equity. As Weale (1992: 213) aptly observes, "no country (...) has discovered how to combine technical effectiveness with political responsiveness and economic efficiency. The solution to that problem still awaits discovery".

One policy instrument which offers an interesting opportunity to achieve both effectiveness and efficiency is emission trading (or marketable permits). The Canadian economist Dales invented this instrument in 1968. The main idea behind this mechanism is that firms with the lowest marginal abatement costs should abate their emissions more than firms with the highest abatement costs. The first steps in an emission trading scheme are, in general, taken by the government which defines the emission levels for a particular region and then fixes an amount of permits which subsequently are either sold to the highest bidders at auctions or distributed for free (so-called grand-fathering). At this stage the government opens up the game for the market forces. The polluters participating in the scheme start to sell and buy their permits. Emission permits will be bought by those firms which have the highest opportunity costs (Barde, 1995).[40] It should be added here that the pros and cons of

[40] The United States has, as the only OECD country, systematically explored the possibilities to involve the market forces in solving environmental problems and introduced marketable permits already in 1976 (OECD, 1994a). An emission trading policy for air pollution was issued in 1982 and allows the trading of emission reduction credits in the following forms (OECD, 1994a: 88): (1) *offsets*, which are tradeoffs that enable new sources to add to the pollution load in polluted areas by getting an old source in the same area to reduce its pollution load to a greater amount; (2) *banking*, which means that industry can "bank" emission rights for later use if a new source was not available at the time an old source was retired; (3) *bubbles*, which mean that sources are given flexibility to meet standards that have been defined on emission point basis; and (4) *netting*, which means that as long as no net increases in emissions occur within a facility, major sources of pollutants can be exempted from new source requirements.

The 1990 acid rain programme in the USA is almost exclusively based on emission trading. It has brought about emissions reductions of sulphur dioxide and other air pollutants in a cost-effective way. Emission trading has also been used in the USA to phase out production of CFCs and leaded petrol. The question arises as to why marketable permits are not more extensively used within the OECD. Possible explanations are the following. First, not all countries use uniform emission limits, which means that there is little incentive to develop the flexibility that a system with marketable permit implies. Second, environmental organizations tend to object the idea that a firm can acquire a right to pollute. Thirdly, transaction costs (for monitoring etc.) are typically higher under emission trading than under a system of uniform emission standards. These costs are likely to be borne by the administration who therefore may be resistant to reform (Weale, 1992).

emission trading have been frequently debated in Poland in the 1990s (see Chapter 6).

The third approach is information and moral suasion aiming at changing an agent's behaviour on a voluntary basis. This could be accomplished via education, transfer of knowledge, training, persuasion, recommendation, and negotiation. One important instrument in this category is voluntary agreements between governmental agencies and private enterprises.[41] This type of policy instrument is likely to gain importance in the future. According to the OECD (1994b: 13), a shift towards prevention and sustainability will require governments to use instruments such as negotiation with stakeholders and joint agreement and action plans between sectoral ministries.

It should be noted that in real life policy instruments tend to come in packages (for example, regulations are almost always followed by some kind of information). Moreover, the application of policy instruments tends to require some kind of organizational arrangements (Vedung, 1995: 15).

Each type of policy instrument has its strengths and weaknesses. As regards regulations, a major advantage is that they are most suited to effectively prevent hazards and irreversible effects. Furthermore, regulations frequently provide polluters with incentives to develop the technology. Provided that there is effective enforcement, these instruments are able to achieve the desired environmental goals. The point is that enforcement is often problematic "(...) mainly owing to the great number of control, administrative requirements, staff (...), legal procedures in case of non-compliance and so on" (Barde, 1995: 207). A second drawback is that command-and-control instruments tend to become weakened by bargaining and negotiation between representatives of the polluters and the environmental authorities. A third dilemma is that regulations are considered to lack incentive for rapid action, for example, development of new technology. Lastly, regulations are expensive for society in that they are often not efficient in economic terms.

Economic instruments, such as environmental taxes and charges, minimize total abatement costs in that they constitute a permanent incentive to reduce pollution.[42] Furthermore, they may provide a source of revenue (Barde, 1995: 211). However, a number of problems and uncertainties arise in connection with the use of these instruments. First of all, the rate of the charge and taxes is not always fixed at a level ensuring that the instruments will become effective in environmental terms. Secondly, charges and taxes are inappropriate for controlling toxic and hazardous substances. The best way to control these substances is by means of direct regulations and bans. Thirdly, there are distributive implications that must be taken into consideration when economic instruments are used. For instance, energy taxes may have negative effects on poorer households.

Also voluntary agreements have their pros and cons. On the one hand they offer flexibility and transparency. On the other hand the environmental authorities' control over actual implementation is minimal.

---

[41] "A reasonably informed public can bring valuable insights into the policy-making and enforcement processes. At the same time, directly involving members of the public in the formation and implementation of policy can lead to much greater public interest in the success of the policies and awareness of the public's stake in the outcome" (OECD, 1994b: 35).

[42] In the ideal case the tax should correspond to the external cost. Such a tax is called a Pigouvian tax after the British economist Arthur C. Pigou. There are virtually no practical applications of Pigouvian taxes. Two exceptions are the sulphur and nitrogen-oxide taxes which were adopted in Sweden in 1991 (Żylicz, 1995b).

2.4. Belief systems

Policy choices contain elements of empirical belief, practical judgement, and ethical conviction (Lindblom, 1980: 117). These beliefs, or policy theories as they also are called, can be reconstructed in order to enhance the understanding of the positions of the actors taking part in the policy-making process. A policy theory is "the *total* of causal and other assumptions underlying a policy" (Hoogerwerf, 1990: 285). These assumptions concern relations between objectives and means (final relations), between causes and effects (causal relations), and between principles and norms mutually or between principles and norms on the one hand and existing or expected situations on the other hand (normative relations). The underlying assumptions of the policy theory may also refer to the whole policy process, the policy organizations, and the particular policy field (*ibid.*: 286). Explicit or implicit policy theories can be found in either written or oral statements made by policy-makers about a particular policy. Hoogerwerf distinguishes three approaches to the evaluation of a policy theory. The first approach is mainly interested in the content of the ideas. The second approach is primarily interested in the patterns of the perceptions. The third one studies the validity of the arguments put forward (*ibid.*: 288). All of these three approaches should be carefully studied in research on policy theories.

Hoogerwerf claims that "the structure and quality of policy theories have effects on the policy contents, the policy process, and the policy results" (*ibid.*: 290). For example, if the policy theories of the policy-makers and the implementers do not converge, implementation may end up with a "fiasco"; "policy failure can be partly accounted for by the policy theory" (*ibid.*).

It can be argued that governments can and must act even in the absence of knowledge about what will work and what will not. Rothstein (1994) argues that there is no need for an established causal theory at hand before a policy is implemented. In his view, the achievement of successful results is often a matter of organizing the implementation process in such a way that the uncertainty in the policy theory can be compensated by flexibility in the implementation stage.

Environmental policy debates can be reconstructed as a struggle between rival policy belief systems (Hoppe and Peterse, 1993: 29). A belief system consists of a set of beliefs (or policy theories as Hoogerwerf calls them) which are held by an actor. A reconstruction of an environmental policy can be made by distinguishing between (a) a policy claim: for example, the problem is a matter of reducing health risks; (b) causal and normative arguments supporting the policy claim, for example: scientific findings on air pollution; (c) a claim about methods and instruments for solving the problem, for example: economic instruments like a product charge; and (d) a condition which is supposed to validate the policy claim, for example: this claim is true if there is consensus about the fact that protecting the people from health risks is the only and principal policy objective (Hisschemöller, 1993).

A number of scholars hold that both policies and belief systems are hierarchically ordered and exhibit a core and a periphery. The core assumptions tend to remain stable over long periods of time while the peripheral beliefs are easy to change (Weale, 1992: 59).

The reconstruction of belief systems is a useful method for analysing public policies. But the method should be handled with some care. Non-policy beliefs can be important factors in explaining policy-making. For example, Loeber and Grin (1996) found that in Dutch eutrophication policy the detergent industry was driven by professional beliefs related to its management paradigm. Furthermore, it sometimes happens that individuals hold conflicting beliefs, values, and attitudes, for example, about the role of the state *vis-à-vis* various problems in society (Myrdal, 1944, 1966). Another point is raised by Weale, who holds that belief systems can be conceived as effects rather than causes, that is, rationalizations for underlying economic and/or political interests. He deserves to be quoted at some length:

(...) although certain belief systems may be essential for the legitimation of public policy strategies and principles, there is no reason to think that the holding of any belief by members of the policy élite is the main cause of subsequent policy outcomes. Thus, it would clearly be too simple to ascribe the long boom of the post-war period to the adoption of Keynesian principles among economic policy elites, rather than to more general structural features of the world economy. The specification of belief systems may be essential to the identification of what policy élites take themselves to be doing, but it may not be a leading part of the explanation of subsequent policy outcomes, or perhaps even policy outputs (Weale, 1992: 58).

2.5. The role of external factors and stable parameters

Environmental policy-making does not take place in a vacuum, it is not a phenomenon which is isolated from other societal processes. On the contrary, it is affected by political, social, technological, and economic events that take place outside the environmental policy-making process. There are numerous examples of how environmental policy directly or indirectly has been shaped by such events. A clear-cut example is the Arab Oil embargo of 1973-74 that led to an upsurge for energy-saving measures in the western world. Economic downturn, high inflation, and high unemployment tend to reduce the electorates' preparedness for economic sacrifices for the sake of environmental protection. Prospering economies have created so-called post-materialist values that have been one of the driving forces for the emergence of Green Parties. Parliamentary and presidential elections may play a crucial role. For example, the election of Ronald Reagan in 1980 led to a sharp cut in the Environmental Protection Agency's (EPA) budget, its staff was changed, and enforcement was obstructed (Lester, 1989: 14). Environmental protection has been heavily promoted in Spain, Portugal, and Greece after these countries' entry to the European Union. For example, some 80 per cent of the present environmental regulation in Portugal has come about as a response to EU directives. In Spain, "the EU has prompted the gradual recognition by industry of the importance of the ecological issue (...)" (Aguilar Fernandez, 1994: 47-48). The environmental expenditures in Spain increased by almost 50 per cent two years after the country had joined the EU (Almarcha Barbado, 1993: 63).

Besides these dynamic factors, there is a set of more stable factors, also external to the environmental policy-making process, that shape national approaches to environmental protection. The literature on national environmental policy styles[43] emphasizes that each country's approach to environmental management is influenced by a whole range of general factors which may explain the differences in the countries' approaches to environmental problems and to the choice of solution

[43] See for example Boehmer-Cristiansen and Skea (1991), Downing and Hanf (1983), Eberg (1997), Enloe (1975), Lundqvist (1980), and Vogel (1986).

strategies.[44] Von Moltke (1993: 97) has shown how geographical factors shape approaches to environmental policy in different member states of the European Union: in the northern countries, water quantity is not a pressing issue while water quality is; the Mediterranean countries tend to transfer water pollution to the sea and worry primarily about water supply. Second, historical factors may explain why countries tend to prefer certain approaches and structures and avoid others. For example, it may be impossible to create a fully integrated environmental agency in the Netherlands because of the fact that control of watercourses and canals is vested in the Ministry of Transport, with responsibility for the historically important *Waterstaat* (Weale, 1992). A third important aspect is the political system and the political culture, which both affect the national environmental policy styles. Vogel's (1986) comparison between environmental regulation in the USA and Great Britain, and Lundqvist's (1980) comparison between clean air policies in the USA and Sweden, both led to the conclusion that the characteristics of a political regime are more important than the nature of the particular policy area. Weale (1992: 55) has pointed out that the room for manoeuvre for the policy-makers is larger in a unitary system than in a federal system. Fourth, economic structure and economic situation play an important role. According to Lundqvist (1980), policy-makers in Sweden and the USA in the 1970s arrived at different conclusions about what environmental policy measures are the adequate ones, due to different importance of export in the two countries. A last factor that may be decisive is the distribution of natural resources. Countries dependent on coal for supply of energy, such as Poland and Australia, may find it more difficult to implement measures to reduce sulphur dioxide and carbon dioxide emissions than countries that rely on other energy sources such as gas, nuclear power, and water power.

2.6. The advocacy coalition framework

The advocacy coalition framework (ACF) is a synthesis of several decades of research devoted to the explanation of policy change. It is a comprehensive framework integrating virtually all relevant factors internal and external to the policy-making process. In this respect the ACF distinguishes itself from other models. Moreover, the ACF has been recognized as a useful ordering device for policy analysis. To this can be added that the ACF has proved to be a fruitful approach to studying and better understanding the policy-making process in at least twenty-nine major studies carried out in the USA and some other western countries (Sabatier and Jenkins-Smith, 1999).

The ACF has four basic premises. Firstly, understanding policy change requires a long-term perspective. According to the framework, it is necessary to study policy developments over at least ten years in order to explain policy change. Secondly, it is most useful to focus the analysis on a *subsystem*:[45] "Subsystem members include those policy elites who, with relative regularity, follow and attempt to influence policy developments in a given issue area" (Sabatier and Jenkins-Smith, 1993: 241).

[44] Including decisions, non-decisions, and agenda-building models.

[45] New subsystems emerge, at least partly, as a result of other subsystems' neglect of a particular problem (Sabatier and Jenkins-Smith, 1993: 24). Subsystems tend to be under continuous political construction. Unstructured policy problems often arise between two or more established subsystems (Grin and Hoppe, 1997: 7).

Thirdly, policy subsystems consist of actors of governmental and non-governmental institutional spheres. Insofar as they are governmental actors, they usually span more than one level of government. Lastly, public policies can be conceptualized as belief systems[46] (*ibid.*: 16).

The ACF is depicted in Figure 2.1. On the right side is a policy subsystem (for example, environmental policy) in which the actual policy-making process occurs. The actors in this process are aggregated into *advocacy coalitions* which consists of "people from a variety of positions (elected and agency officials, interest group leaders, researchers etc.) who share a particular belief system – that is, a set of basic values, causal assumptions, and problem perceptions – and who show a nontrivial degree of coordinated activity over time" (*ibid.*: 25). It should be emphasized that not all actors in a particular subsystem are members of a coalition. Some actors appear as *policy brokers* whose main aim is to search for compromises between the coalitions (*ibid.*). Other actors, for example, some researchers and bureaucrats, are not primarily interested in achieving policy change but participate in the policy-making process only because they have certain skills to offer (*ibid.*: 27).

The *belief systems* of the advocacy coalitions are structured into three different categories (Sabatier, 1998: 112-113):

1. A *Deep Core* of normative and ontological axioms that is part of the personal philosophy and which apply to all policy areas. Deep core beliefs are thus very difficult to change.
2. A *Policy Core* of basic strategies for achieving deep core beliefs in a particular subsystem. Central policy core items are: orientation on basic value priorities (for instance, the proper relationship between environment and economy), identification of groups or other entities whose welfare is of greatest concern, overall seriousness of the problem, basic causes of the problem, the role of the government vs. the market, proper distribution of authority among levels of government, priority accorded various policy instruments, method of financing, ability of society to solve the problems, and participation of the public vs. experts and elected officials. Policy core beliefs are not easy to change but changes do occur. It should be emphasized that it is the policy core beliefs that are decisive for the scope of subsystems and the formation of advocacy coalitions.
3. A set of *Secondary Aspects* comprising instrumental decisions necessary to implement the policy core in a specific policy area. These aspects are moderately easy to change and are the topic of most official policy-making.

The main goal of coalitions is to translate their beliefs into legislation or administrative agency rules, referred to as *policy output* in the policy subsystem box. The policy output leads to *policy impacts,* which is "the ability of governmental institutions to accomplish policy objectives" (Sabatier and Jenkins-Smith, 1993: 2).

The two boxes on the left side of Figure 2.1 are entitled *external (system) events* and *relatively stable parameters*, respectively. External events refer to short-term socioeconomic conditions, changes in public opinion, changes of system-wide governing coalitions (caused by, for instance, elections), and impacts from other subsystems. There are four categories of relatively stable parameters: basic attributes of the problem area, basic distribution of natural resources, fundamental socio-cultural

---

[46] It is assumed that beliefs constitute "the principal glue of politics" (Sabatier and Jenkins-Smith, 1993: 27).

values and social structure, and basic constitutional structure (*ibid.*: 224). The external events and the stable parameters create *constraints and resources* for the subsystem actors (see the small box in the middle of Figure 2.1).

According to the ACF, there are three causal drivers for major policy change (that is, changes in the policy core elements in a legislative text or a public programme): (1) The competition between different advocacy coalitions. (2) The impact of stable parameters on constraints and resources of the actors within and between the coalitions. (3) The impact of external events on constraints and resources of the actors within and between the coalitions. It is the external events that are the "principal dynamic elements affecting policy change" (*ibid.*: 22).

To improve their possibilities to influence public policy, Sabatier hypothesizes, advocacy coalitions engage themselves in *policy-oriented learning,* which is defined as relatively enduring alterations of thought or behavioural intentions that result from experience (*ibid.*: 42). Policy-oriented learning involves, among other things, increased knowledge about the policy problems and perceptions concerning external events (*ibid.*: 19). Learning can be the result of processes such as individual learning, the diffusion of new beliefs, turnover of individuals, group dynamics, and patterns of communication. Learning is most often used in an instrumental way, that is, the coalitions want to understand the world in order to further their policy objectives. Instrumental learning relates to changes in the secondary aspects of a belief system. It is assumed that learning may occur within a coalition's system as well as across belief systems (*ibid.*: 48). Although policy-oriented learning can lead to conceptual policy change (that is, changes in the policy core of a belief system) it is most likely to have an impact on the secondary aspects of a belief system (*ibid.*: 5).

Using the ACF for analysing long-term shifts in environmental policy in a country that has moved from a socialist regime context to a market-based democracy poses several challenges, both for the analyst and the framework itself. A central question that arises is: should systemic change be seen as an external event or a change in a stable parameter? The answer must be the latter because systemic change is a change in the basic constitutional structure. The difference between external events and relatively stable parameters is *not* that the former change and the latter are stable. The difference is the rate of change, like the weather (an external event) compared to the global climate (a hitherto relatively stable parameter).

The socialist framework that existed in Poland until 1989 implied that the policy actors in, for example, the environmental policy subsystem operated under fundamentally different conditions than in the West. The ACF has been developed to explain policy change in a pluralist setting in western polyarchies. The policy-making process in the socialist states of the Soviet bloc was fundamentally different from the pluralist one in the sense that one party controlled the political arena, there was no free press and free debate, and public participation was severely restricted. However, the policy-making process in the Soviet bloc was not completely different from the one in western states in all respects: the ministries tended to act as interest groups (Nove, 1980: 65) and bureaucratic groups did bargain with each other (Dittmer, 1983). Certain politically sensitive issues were not able to be raised in the socialist system because they were deemed unacceptable by the state (Ziegler, 1987: 48).[47]

---

[47] A reconstruction of belief systems from the pre-1989 period is more difficult than in the post-1989 period because under socialism certain beliefs were not allowed to be published. As De Bardeleben (1985: 75) has aptly observed: "In the east bloc there was a censored media. Hence the assumption that people thought thoughts, which never appeared in print". The environmental journalist

These factors certainly have implications for the analysis. It can be concluded that the environmental policy actors under socialism were confronted with a number of constraints not existing in the systemic framework in which the ACF was developed.

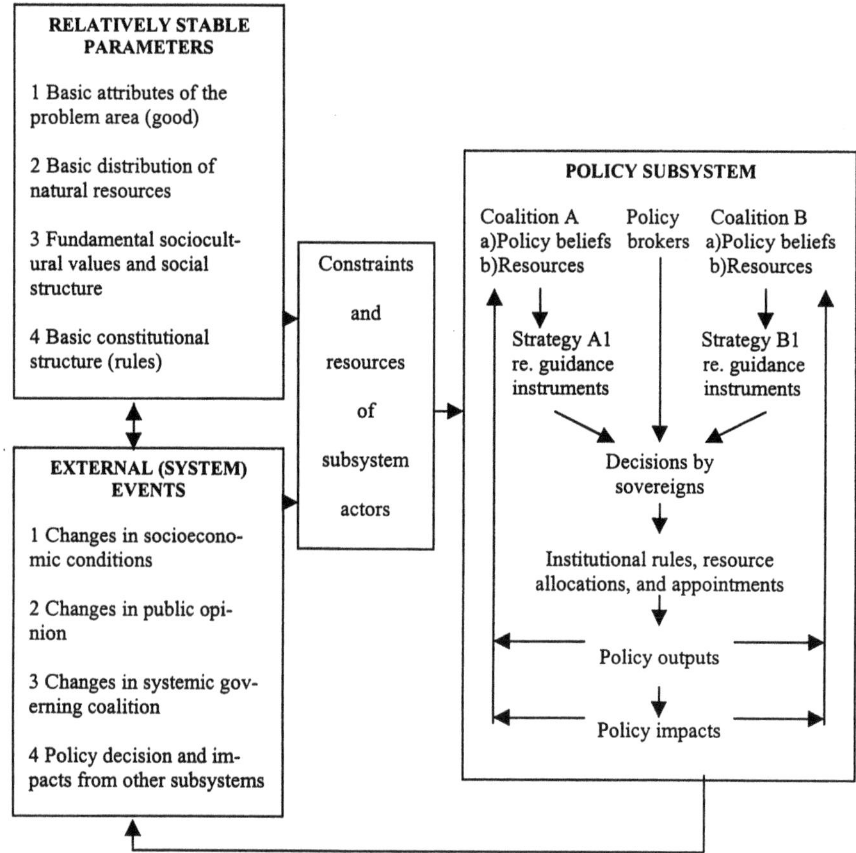

Figure 2.1 The advocacy coalition framework. Source: Sabatier, 1998.

2.7. Research questions and structure of the book

The general research problem dealt with in this book, as already mentioned in Chapter 1, is:

To what extent has systemic change in Poland affected environmental policy in the 1990s?

According to hypothesis 5 of the advocacy coalition framework, "significant perturbations external to the subsystem (e.g. changes in socio-economic conditions,

---

Iwona Jacyna (personal communication, 1995) adds another dimension to this problem in arguing that there was an "inner censorship" at work.

public opinion, system-wide governing coalition, or policy outputs from other subsystems) are a necessary, but not sufficient, cause of change in the policy core attributes of a governmental programme" (Sabatier: 1998: 118). One may thus expect that external events will play an important role in policy change, both before and after systemic change. One may also expect that systemic change as a change in a relatively stable parameter has had a profound impact on the environmental policy subsystem.

Within the environmental policy subsystem, policy change will be measured into four related aspects:

1. Characteristics of the policy process. One may expect that, as a result of the transition to democracy, the setting of the environmental policy agenda has become less of the kind that Cobb *et al.* (1976) refer to as the inside access model or the mobilization model and more of the outside initiative model. One may also expect that political change has brought about a political process and patterns of political interaction that are more pluralistic and less corporatist. In consequence, the period from 1989 onwards will show more opportunities for advocacy coalitions to openly mobilize support for their position, to bargain and trade majority policy options among each other in the environmental subsystem and with stakeholders in related subsystems.

2. Policy content. One may expect a shift in policy content following the fall of state socialism in 1989. Given the general orientation of the Polish policy-making elite towards the west, one may expect that environmental policy is brought in closer conjugation with policies in the OECD and EU member states. Policy content refers to (a) goals, as highlighted in key policy documents and laws, (b) policy instruments (including regulations, economic instruments, and information/voluntary agreements), and (c) the institutional framework designed to carry out environmental policies in an effective manner.

3. Policy implementation and policy impacts. One may expect an improved environmental performance in Poland after systemic change due to environmental policy reform and the emergence of a new economic and political setting.

4. Policy belief systems. As a result of systemic change, one may expect a major shift in the deep core and policy core of the dominating belief system. One may also expect the emergence of new belief systems as a response to the new challenges represented by the transition to democracy and market economy.

In order to answer the key research question, the study will address three subquestions.

1. What environmental policy has developed before and after systemic change?

Related questions are:

On the policy process: Who participated? Which were the dominating issues? Who brought up the issues? How did the actors in the environmental policy subsystem interact? What kind of interactions is discernible between the environmental policy subsystem and various economic and industrial subsystems?

On policy content: What have been the most important laws and policy documents? Which policy goals and policy instruments have dominated? What did the institutional framework look like?

On implementation and policy impacts: How and to what degree was environmental

policy implemented? What policy impacts are discernible?
On belief systems: What advocacy coalitions have been discernible and what were their belief systems? The study addresses for each belief system the deep core, the policy core, and the secondary aspects, respectively.

2. To what extent was policy change radical or incremental?
In answering this question, the study will distinguish between policy process, policy content, implementation and impacts, and belief systems. Related questions thus become:

To what extent has the policy process changed in terms of agenda-setting and patterns of interactions between the environmental policy actors? The change in the policy process is radical if we see a change from one agenda-setting model to another and/or the emergence of new forms of political interaction.

To what extent has the policy content – especially goals, policy instruments, and institutional framework – changed? A radical change in terms of policy content implies new goals, the introduction of new policy instruments, and the creation of new institutions.

To what extent has implementation improved? To what extent have concrete policy effects been achieved? The radicality of the changes related to implementation will be measured by looking at changes in the gap between policy goals and policy performance.

Were there changes in the belief systems? If so, to what extent were these changes observable in the deep core, policy core, and secondary aspects of the belief systems? Radical change in the belief systems implies the emergence of new advocacy coalitions and/or changes in the deep core and policy core of the belief systems of already existing advocacy coalitions.

In addition, this study is also interested in the following questions: Have the four policy categories (process, content, implementation, and belief systems) been changed to an equal extent? Was change most radical in the 1980s or in the 1990s?

3. What have been the main driving forces for policy change and continuity?

Related questions are: How have external events and relatively stable parameters shaped environmental policy? Has learning occurred within or across advocacy coalitions or both? What role has policy-oriented learning played for policy change? Can interaction and competition between advocacy coalitions explain policy change?

Structure of the book

Chapters 3-5 analyse the development of environmental policy in Poland until 1997. The period is divided into three episodes: national environmental policy before 1980, with a focus on the 1970s (Chapter 3), national environmental policy in the 1980s (Chapter 4), and national environmental policy in the 1990s (Chapter 5). A reason for this periodization is that these episodes represent three distinct phases in environmental policy terms as well as in economic and political terms. Each episode starts with an introduction containing a short chronology of the main political and economic events. In this way the key external events of each period are dealt with. This is followed by an analysis from a policy process perspective. Subsequently, the main content of environmental policy is presented. Implementation and policy effects are discussed in sections entitled implementation. Next, the main actors are structured into advocacy coalitions on the basis of their belief systems. Attention is also focused n policy-oriented learning. The last sections summarize the chapters and draw onclusions about policy change.

Next there follow two chapters on sectoral developments. Chapter 6 is devoted to air ollution from stationary emission sources. It first gives a general background to the roblem of air pollution. It then deals with policy adoption and interaction processes. lext, there is a section about four major policy issues in the air protection policy: tandards and permits, emission trading, enforcement, and fees and financing. Then nsue sections on implementation, advocacy coalitions and belief systems, and policy-riented learning. The last section draws conclusions and summarizes the chapter.

Chapter 7 deals with the hard coal sector's discharges of saline waste water, a roblem that has remained unsolved in the 1990s. Chapter 7 has the same structure as 'hapter 6 with one exception: the issue of technology has been given a separate ubsection.

Subquestion 1 will be answered in Chapters 3, 4, 5, 6, and 7.

Chapter 8 answers the key research question and subquestions 2 and 3. It compares ie 1980s and 1990s, both with respect to national and sectoral policies. The chapter efines the impact of systemic change on the policy process, policy content, nplementation/impacts, and belief systems.

.8. Research design and approach

'his thesis uses a case study approach. Case studies are extensively used in sciences uch as economics, sociology, and political science. Case studies are specially suited) research in which how or why questions are being posed. The need for case studies arises out of the desire to understand complex social phenomena" (Yin, 1994: 3).

Yin (*ibid.*: 38) distinguishes four case study designs: (a) single-case (holistic) esigns; (b) single-case (embedded) designs; (c) multiple-case (holistic) designs; and 1) multiple-case (embedded) designs. This study has an embedded single-case esign. The reason for this approach is that only this type of case allows for a time eries analysis of policy changes in the investigated period consisting of repeated bservations of the dependent variable (policy changes) both before and after the ıdependent variable which is systemic change. In order to create some variation in ıe case to control, at least marginally, better for the causal link in focus (the impact f systemic change on environmental policy) the single case (Poland) is split into ıree subcases: policy change at the national level and in two sectoral policy fields, *iz.* air pollution from stationary sources and saline waste water discharges from the ard coal sector. These issues are justified on the grounds that (1) they represent one uccess and one failure in terms of policy implementation; (2) both air pollution and aline waste water have been on the environmental agenda at least since the mid-980s (hence it is possible to take a long-term perspective on these policy fields /hich is a requirement for applying the ACF); and (3) the two subcases highlight, in ifferent ways, the obstacles for environmental policy innovation during the transition eriod. Despite the fact that the policy fields dealt with in the subcases are not ıdependent from the national subcase, these cases are meant to "test" the plausibility f the conclusions drawn about the developments on the national level. Thus, the ectoral subcases are aimed at illustrating various conditions under which the causal nks present on the national level vary in direction and strength.

This study analyses environmental policy developments in Poland by comparing (1) evelopments at the national level in time (1970s, 1980s, and 1990s); (2)

developments at the sectoral level in time; (3) similarities and differences between the sectors; and (4) similarities and differences between the national and sectoral level.

In order to be able to establish a causal link between systemic change and policy change (or the lack thereof) it is necessary to use an analytical framework which lists *all* possible causes for policy change. Hence the selection of the advocacy coalition framework (ACF) as the main analytical framework of this study.

Since the study includes an analysis of interactions between actors involved in the policy process I find it useful to draw on the literature on agenda-setting (Cobb and Elder, 1983). This approach is particularly useful to highlight the interaction patterns of the policy actors. Agenda-setting is an integrated part of the ACF (Sabatier and Jenkins-Smith, 1993), but this stage of the policy process has so far been devoted relatively little attention by the ACF-community. My aim here is to put somewhat more emphasis on agenda patterns than usually is done in studies based on the ACF.

The findings of this research have been reached by making use of multiple sources of information. Firstly, it is based on content analysis of governmental policy papers, memoranda, laws, professional journals, popular magazines, and newspaper accounts. Secondly, both structured and open-ended interviews with more than sixty representatives of central and regional governmental agencies, the parliament, industry (especially the energy and hard coal sectors), the scientific community, environmental organizations, and media have been carried out in Poland, mainly in 1995 and 1996. (See list of the interviewees at the end of the book.) Among the interviewees were five former Ministers of Environment (Waldemar Michna, Józef Kozioł, Bronisław Kamiński, Maciej Nowicki, and Stefan Kozłowski) together with Jan Szyszko who held this post when this book went to the printing office. In certain instances, an interviewee is a representative of two or more "estates". The most striking example is that many NGO activists have a background in the scientific sector.

The respondents were asked questions about, *inter alia*, problem perceptions, policy preferences, and policy events (see appendix for a list of the main questions asked in Poland). A number of interviewees served as informants by providing advice about selection of the interviewees (according to the so-called snowball technique) and relevant background information.

2.9. Summary

This chapter has involved a search for an appropriate policy framework in order to investigate long-term changes in Polish environmental policy. In doing so it examines existing policy-making theories briefly and selects the advocacy coalition framework. The rationale for this is that the ACF aims at providing a comprehensive and dynamic explanation of why and how long-term changes in public policy occur. The chapter also presents the research questions and the research design and approach.

PART II

NATIONAL POLICY

3. SETTING THE STAGE

3.1. Introduction

This chapter has the following structure. A first step to operationalize the ACF is taken in Section 3.2 which identifies the key features of the relatively stable parameters in Poland. Section 3.3 briefly presents the main external events in Poland betwen 1945 and 1980. Section 3.4 investigates the earliest roots of environmental protection in Poland. Section 3.5 addresses the main features of the policy process in the 1970s. Section 3.6 focuses attention to the main environmental policy documents of the 1970s. Section 3.7 is about implementation of environmental policy in the 1970s. Section 3.8 reconstructs the dominating belief system up to the 1970s and discusses policy-oriented learning. Section 3.9 draws conclusions about driving forces for policy change and briefly summarizes the chapter.

3.2. Relatively stable parameters in Poland

Fundamental cultural values and social structure

Poland's geographical location between imperial Germany/Hitler's *Reich* and Tsarist Russia/the Soviet Union has been unfortunate for the Polish state and the Polish people. Poland's history is full of wars with these neighbours. Poland disappeared from the European map after three successive partitions of the country in the late eighteenth century (1773, 1793, and 1795). Eastern Poland fell under Russian rule, southern Poland fell under the Austrian empire while the western part came under German control. The foreign occupation would last until 1918 when Poland regained independence (Davies, 1986).

Almost 15 per cent of the Poles lost their lifes in the first world war. In the second world war, in which Poland was attacked by both Nazi Germany and the Soviet Union, six million Polish citizens were killed (*ibid.*).[48]

Poland's history has produced a mentality where the preservation of national identity and the restoration of national independence are central elements (*ibid.*: 200). Patriotism is a vital element in Polish consciousness and Poland's Westernism is fundamental. It is a general perception in Poland that the country is a part of central Europe rather than eastern Europe (*ibid.*: 343).

A typical feature in Polish history, at least since the latter part of the sixteenth century, is the resistance against central power *of any kind* and a deep belief in personal freedom. The strong belief in individual freedom was epitomized by *liberum veto*, which meant that one single member of the nobility could reject the legislation

[48] The Polish casualty rate in the second world war was 18 per cent. It was a casualty rate which was considerably higher than in any other country involved in the war: the Soviet Union: 11.2 per cent, Yugoslavia: 11.2 per cent, Germany: 7.4 per cent, Japan: 2.5 per cent, United Kingdom: 0.9 per cent, and the United States: 0.2 per cent (Davies, 1986: 64).

of the parliament (Davies, 1994: 427-448). This system succeeded in weakening monarchial despotism but it also prevented effective collective action.

These observations are confirmed by the fact that the periods of full independence (1918-39 and 1989 and onwards) have been characterized by weak governments and fractious parliaments. In 1926 there were more than thirty parties elected to the parliament and since 1989 Polish politics have been plagued by fragile multiparty coalitions and weak executive power (Sachs, 1994: 112-113).

According to the Polish sociologist Maria Marody, Poles tend to see politics as a sphere of activity beyond the reach of ordinary people.[49] People see politics as a "product" not as a process, as the outcome of decisions made elsewhere and over which the ordinary person neither has nor expects to have much control (Batt, 1991: 45). The divisions between rulers and ruled, between "us" and "them", were apparent in the socialist period (Millard, 1994: 27) and continue to characterize Polish political culture.

Poland is, together with Ireland, the most catholic country in Europe (Davies, 1986: 342). Traditional catholic values are adhered to by a large segment of the population.

During the socialist period the Polish political discourse (both the official and the non-official) had a left-right dimension. In the 1990s a new political split has been added to this: there are those who want to see Poland as a secular and modern European state while there are others for whom Poland should remain adherent to traditional catholic and national values (Węcławowicz, 1996: 92).

The dominating economic sectors in the 1990s in terms of employment are the service sector, industry, and agriculture. The trend in the 1990s is that people leave agriculture and industry for the service sector (Nowicki, 1993: 177). Polish agriculture is small scale and inefficient. At the end of the 1980s there were 2.7 million private farms with an average size of 7.2 hectares and 30 per cent of farms had less than two hectares (Balcerowicz, 1995: 316). Industrial production has traditionally been dominated by heavy industry including mining and production of iron, steel, and chemicals (OECD, 1995a: 17). In the 1990s, food processing and light industries have become increasingly important industrial sectors.

The Polish environmental consciousness is deeply rooted in the country's culture and history. Since the latter part of nineteenth century nature is

> (...) filtered by our perceptions of how the world should look. The birch forest is much more interesting for us if there are partisan graves in it. An untouched forest is much more worth protecting when we can associate it with different historical values. Programmes for the protection of oaks tend to be motivated by their historical value, not because of their role in the nature.[50]

In a similar vein, Karaczun and Indeka (1996) argue that nature areas of importance for Poland's history and culture tend to be protected.

Despite the fact that Poland has a long coast along the Baltic Sea, Poles tend not to identify themselves with the Baltic Sea region. This may have an historical explanation: before the new borders were established in 1945, Poland was an inland country. Generally speaking, Polish architecture has neglected the water perspective.

---

[49] In one opinion poll conducted in 1993 over half of the respondents accepted the statement that "nobody needs political parties except their leaders and activists" (Cirtautas, 1997: 243).

[50] Personal communication with Witold Maciejewski, 1995.

"In Poland, it has not been common to look at the water from where you live. And *vice versa*: it is not relevant how one's house looks like from the water perspective".[51]

Basic distribution of natural resources

Poland is mainly a lowland country.[52] The great Polish plain stretches from the German border along the Odra river in the west to the borders with Lithuania, Belarus, and Ukraine in the east. Southern Poland is formed by a hilly upland and along the southern border there are two mountain ranges, the Sudety and the Carpathians. A total of 99.7 per cent of Poland lies in the Baltic Sea Basin. Poland is poorly endowed with freshwater resources. Available water resources amount to some 1,500 cubic metres per year, one of the lowest ratios in Europe (OECD, 1995a).

Poland is rich in minerals. It has large reserves of hard coal and is one of the most coal-dependent countries in the world. Without access to coal, it is unlikely that Poland's early industrialization in the nineteenth century could have taken place, let alone the socialist shock industrialization after the second world war. In the three decades during the post-war period coal production, mainly located in Poland's industrial heartland in Upper Silesia, more than doubled. Already before the annual production peaked at 201 million tonnes in 1979, Poland had become the most important coal producer in Europe (the former Soviet Union excluded). Also after the transformation to a market-based democracy, Poland has continued to rely on coal. In 1992 approximately two thirds of Poland's total primary energy supply was covered by domestic coal (Radetzki, 1995: 7).[53] The importance of hard coal for the Polish economy is also underscored by the fact that it has traditionally been an important export product.

Besides hard coal there is brown coal in central and western Poland, copper ores in the west (Legnica and Głogów), sulphur ores in Tarnobrzeg in central Poland, salt in three regions, and non-ferrous metals (zinc and lead ores) in Upper Silesia. The most important mineral deposits are located in the southern and western parts of Poland, which explains why most of the industrial development has been located there (Nowicki, 1993: 178-179). Consequently, the most serious environmental problems appear in these parts of the country.

Basic legal structure

Poland's first constitution, the second oldest in the world after the American constitution, was laid down in 1791 (Davies, 1986). Between 1795 and 1918 Polish territory was controlled by Germany, Russia, and Austria, which imposed their constitutions on the occupied areas.

Soviet domination in Poland after world war two resulted in the establishment of an authoritarian political system centred around the communist party and an economic system based on central planning and state ownership (Balcerowicz, 1995: 290). A

---

[51] Personal communication with Witold Maciejewski, 1995.

[52] The name of Poland (*Polska*) refers to one Slavonic tribe, the *Polanie* ("the people of the open fields") which in pre-historic times settled in the vicinity of modern Poznań (Davies, 1986: 283).

[53] In 1993, coal contributed to more than 80 per cent of the heat that was produced in Poland (Karbownik, 1995). For a comparison: in 1993 coal's share of the primary energy market within the European Union was 13 per cent (Parker, 1994).

socialist constitution was approved in 1952 which confirmed the leading role of the Party and its nomenklatura (Davies, 1996). Two paragraphs about environmental protection were added to the constitution in 1976 (Kabala, 1988).

Figure 3.1 Map of Poland.

Since 1989 Poland has been a democratic state where the legislative power rests in the parliament. The parliament has two chambers, *Sejm* and *Senat*. Parliamentary elections are organized every four years. The Polish President is elected every six years in a general election. Between 1975 and 1998 Poland was divided into forty-nine provinces (voivodships or w*ojewództwa*). Since 1 January 1999 the number of provinces is sixteen. The executive at the province level is the voivod (*wojewoda*) who is appointed by the government (Nowicki, 1993). Presently there are 2,460 municipalities. During socialism the municipalities were completely subordinated to the central state authorities. This was changed with the 1990 Local Government Act that made the municipalities free from direct state control (Węcławowicz, 1996: 173).

After a referendum in 1997, a democratic constitution was approved. The new constitution contains a number of provisions that relate to environmental protection. Most importantly, Article 5 makes sustainable development a principal task of the state and article 74 requires authorities to support the public in its efforts to protect the environment (Jendrośka, 1997: 106).

3.3. Main external events 1945-80

Until the early 1950s Poland remained a typical agrarian nation with relatively minor pollution problems, at least compared to the more industrialized countries of western Europe. However, this does not mean that Poland was spared any environmental damage before it was subjected to socialism. Although the industrial impact on the environment was relatively small in the prewar period, air and water pollution occurred in areas such as Łódź (textile industry) and Kielce-Mielec (metallurgical industry).[54]

In the 1950s, Poland embarked on a rapid transformation from an agrarian nation to an industrial one. This forced industrialization, along the lines of the Stalinist development paradigm, was biased towards heavy industry.

Figure 3.2 Polish bank-note from the 1950s.

The first industrialization plan between 1950 and 1956 envisaged the construction of more than 1,200 new industrial plants (Węcławowicz, 1996: 33). Production of steel, chemicals, and various metals increased substantially. An astonishing example

[54] The first industrialization began in Upper Silesia, then belonging to Prussia, in the early nineteenth century (Sachs, 1994: 11). The first steam pump in this area was installed as early as 1788 in the mining town of Tarnowskie Góry (Manser, 1994).

is that the production of copper increased from 50,000 tonnes in 1950 to 6.4 million tonnes in 1970 (Chmielewski, 1988: 55). The use of indigenous coal, often low-grade with high sulphur content, came to assume ever greater proportions (Juhasz and Ragno, 1993).

In 1970,[55] Edward Gierek replaced Władysław Gomułka as the first secretary of the Communist party (Polish United Workers Party, *Polska Zjednoczona Partia Robotnicza* or PZPR). Gierek greatly expanded Poland's ties with western industrialized democracies in an attempt to obtain western credits for the further industrialization and modernization of the country. Sachs (1994: 26-27) explains:

> The main idea was to stimulate new exports by importing modern technology, while remaining squarely within the socialist framework. Between 1970 and 1977, Poland borrowed about $20 billion from Western governments and banks. The idea was simple, though wrong: that Poland did not need to change the economic system, but just to use better machinery. Remarkably, as with Argentina at around the same time, the huge amount of foreign borrowing produced almost no increase in Poland's exports to Western markets. As a result, the loans could not be repaid. Poland became one of the first countries in the world to fall into debt crisis after a round of heavy borrowing in the 1970s.

Only a third of the loans taken in the 1970s were used for industrial modernization. The remaining parts were spent on import of raw materials and consumer goods (Górka, 1991). No more western loans were granted Poland after 1978. In 1988 the debt stood at $45 billion. Poland's external debt consumed more than one third of the export earnings over the years 1982-86 (World Bank, 1989a).

3.4. Environmental protection before 1970

Poland's first effort in modern times to protect the environment was the 1922 Water Law which regulated, through the introduction of permits, the discharge of industrial effluents into water bodies.[56] It also introduced economic instruments on a limited scale (Cole, 1995a: 300-301).[57] The newborn Polish state had high ambitions with respect to nature protection. A State Council for Nature Protection (*Państwowa Rada Ochrony Przyrody* or PROP) was set up in 1925 and a Nature Protection Act[58] was

[55] After the tragedy in Gdańsk in 1970, when the state's armed forces killed several workers demonstrating for higher salaries.

[56] Poland has an animal protection tradition that emerged long before the nineteenth century. In the eleventh century the king Bolesław Chrobry prohibited general hunting on the beaver which was already rare then. In 1523 a law was proclaimed which had the aim of protecting rare animals such as wisent (European bison), aurochs, beaver, falcon, and swan. In 1868 a nature protection act was passed by the parliament in Lwów. (At this time, Lwów was the capital of the Austro-Hungarian province Galicia which comprised some parts of southern Poland and western Ukraine.) This may be the world's oldest nature protection act issued by a parliament (Chmielewski, 1988).

[57] "(...) the administrator could require compensation for any resulting water pollution damages or assess a fee for the mere privilege of using the public's waters for the 'non-ordinary' purpose of discharging effluents. These fines and fees created, at least potentially, an incentive to conserve water quality and quantity" (Cole, 1995a: 301).

[58] This Act envisaged the establishment of a nature protection fund and a nature protection police force. However, these ideas were never implemented.

adopted in 1934. In the inter-war period, seven national parks were established (Chmielewski, 1988: 24). Moreover, PROP organized 180 nature reserves and approximately 4,500 natural monuments. In 1933 nature conservation became a mandatory part of the secondary school curriculum (Cole, 1995a: 302-304) and Poland's first environmental organization, the League for Nature Protection, was created in 1928 (Kozłowski, 1994).[59]

In 1939 Poland's ambitions to protect its environment were interrupted by the second world war. Nazi Germany's forces murdered many of the leaders of the nature conservation movement. One of them was professor Michał Siedlecki, a zoologist from the Jagellonian University in Kraków and a pioneer of the Polish biological investigations of the Baltic Sea, who died in the concentration camp in Sachsenhausen in 1940 (Sikora, 1988: 185).

The strong conservationist tradition that had been established in the 1920s and the 1930s, survived the war. PROP was revived, and in 1949 the Polish parliament enacted a new Nature Protection Act. It covered natural monuments, nature reserves, national parks, and various endangered species of plants and animals. Between 1949 and 1960 nine new national parks, several nature reserves, and a large number of monuments were established (Cole, 1995a: 309).

In the mid-1950s when the first stage of shock-industrialization had been completed, a considerable level of air and water pollution was present in many places, especially in Upper Silesia in southwestern Poland. However, pollution was not perceived as a problem by the policy-makers.[60] Instead their attention was focused on water management. They were well aware of the fact that Poland was poorly endowed with freshwater resources and that the lack of drinking water for the population and process water for industry constituted an imminent danger, a hindrance for further urbanization and industrialization of the country. In other words, the lack of water was perceived as a developmental barrier and was, hence, taken seriously. As Waldemar Michna, Minister of Environment in 1987-88, puts it: "we were forced to take action".[61] As a result of this awareness the Central Water Management Board (*Centralny Urząd Gospodarki Wodnej* or CUGW) was created in 1960. Its main task was to prepare and administer long-term water management plans.[62] It was a powerful body, gathering the best specialists in the country under one umbrella. It had the legal authority to coordinate activities of other state agencies relating to water management. Thanks to its activities, Poland, at least temporarily, managed to overcome (by, for example, building large water reservoirs) its developmental barrier caused by the scarcity of water resources. CUGW was inexplicably dissolved in 1972 but is still considered to have been an effective water management body.

In the early 1960s Poland took its first steps towards a more comprehensive

---

[59] However, from a western perspective, Poland's efforts to protect nature occurred comparatively late. National parks were established in the USA in 1872 (Yellowstone), Australia in 1879 (Royal National Park), and Canada in 1885 (Banff National Park) (McCormick, 1992).

[60] It should be noted that already in 1954 the Sanitary Epidemiological Service (SANEPID) had been created. It was, *inter alia,* in charge of monitoring of drinking water quality and ambient air quality (MGTiOŚ, 1975: 4).

[61] Personal communication, 1995.

[62] It also had a mandate to control water pollution (Cole, 1995a: 313).

approach to environmental protection. The *Sejm* (the Polish parliament) adopted a new Land-Use Planning Law in 1961 and a new general Water Law in 1962. The latter introduced criminal sanctions, such as fines and imprisonment, for serious violations of the permits. In 1964 a new Civil Code was enacted by the *Sejm* enabling individuals to take legal action to eliminate environmental pollution. Poland's first Air Protection Law was enacted in 1966, introducing the concept of "permissible concentrations" and giving health inspectors the right to close an industrial plant if its emissions posed a threat to human life. In connection with this law, an air protection office was established within the CUGW (Cole, 1995a: 315-316). The law also introduced fines for air pollution (Alberski, 1996).

In the 1960s high chimneys were beginning to be built to disperse air pollution in order to avoid high air pollution concentrations locally.[63] The first environmental protection departments in industrial plants were created in the mid-1960s (Górka and Poskrobko, 1987: 99). In the first decades after the second world war the role of the scientific community became increasingly important. A central figure was Walerian Goetel, a biologist based in Kraków. In the 1960s he made an attempt to create a new scientific discipline, sozology (*sozologia*), for the purpose of protecting the environment. He derived the name from the Greek word *sodzo* which means to protect and to save (Tobera, 1988: 183). Goetel's ideas were never realized but his work was an important source of inspiration for many scientists and students who later on would form Poland's independent environmental movement.

To summarize, the main features of environmental protection in the People's Republic of Poland before 1970 were nature conservation and water management. The former was inherited from the traditions established in the inter-war period, the latter was a necessity for further industrial development. Both issues were relatively well understood and managed. In addition, initial legislative steps to protect water and air were taken, but the notion of "environmental policy" was not yet known.

3.5. The policy process in the 1970s

Environmental policy in the 1970s was determined by a small group of people. Environmental protection was seen as a sectoral task that could be delegated to engineers, biologists, and other experts. The period was dominated by policy initiatives coming from the communist party and the government. There was almost no public pressure on the politicians and the policy-makers, and no alternatives to the official policy plans were presented.

There was little interaction between the environmental policy-makers and the economic sectors. Industry did not bother to influence the content of the environmental regulations. Any policy could easily be blocked by industry in the implementation stage by simply ignoring it. Clearly, environmental policy was made and implemented in a state corporatist manner.

Several events in the West spurred the promotion of environmental policy in the early 1970s: the enactment of the National Environmental Policy Act in the USA in

[63] Edmund Potok, the chief electrician at the Łaziska smelter in Łaziska Górne in Upper Silesia between 1955 and 1970, recalls an interesting episode from 1960. The Party secretary at the plant had been informed by an engineer that the new chimney was 150 metres high and would disperse the air pollution over a distance of 600 kilometres, and made the following comment: "Thus, we shall drown the imperialists in dust" (personal communication with Edmund Potok, 1998).

1969; the publication of the so-called U Thant report in 1969;[64] the creation of an environmental committee within the OECD in 1970; the UN Conference on the Human Environment in Stockholm in 1972,[65] and the adoption of the first environmental action programme within the European Communities (1973) (Liefferink, 1995). The importance of international environmental trends in Poland in the 1970s is illustrated by the fact that, by the end of the decade, Poland had signed approximately twenty-five environmental conventions (Aura, 1979a). One of these conventions, the 1973 Gdańsk convention on the protection of the fish resources in Baltic Sea, came about as a result of a proposal from Poland (Sikora, 1988: 154). Furthermore, Poland was an active member of the United Nations Environmental Programme (UNEP) in the 1970s. Poland was represented in its managing council between 1973 and 1978 and the UNEP frequently engaged Polish experts in various projects (Ochocki, 1985: 24).[66]

In the latter part of the 1970s the environment gradually lost its privileged position on the political agenda. Environmental policy reform slowed down and expenditure was sharply reduced. At the end of the 1970s there was a growing awareness of Poland's low environmental performance. Some researchers and professional experts began to prepare reports, articles, and letters (individually or in small groups, not as representatives of organizations) informing the decision-makers about the neglect and calling for a new approach to environmental protection.[67] Complaints were articulated and became more and more common, particularly by the end of the decade. However, the *demands* for change were not elevated to particular *issues*.

3.6. Policy adoption in the 1970s

The Polish policy elite's will to promote the idea of environmental protection in the 1970s was indicated by three important initiatives: (1) the elevation of environmental protection to ministerial level; (2) the elaboration of a long-term environmental programme until 1990; and (3) the preparation of a comprehensive environmental law. The first task was carried out in 1972, when the government established a Ministry of Territorial Management and Environmental Protection (*Ministerstwo*

---

[64] The report "Man and his environment" by U Thant, the then general secretary of the United Nations, appeared in Polish in 1971 (Tobera, 1988: 65). It was positively received by the authorities and the general public and had a great impact on environmental policy development in Poland.

[65] The formulations in two paragraphs about the environment, which were added to the constitution in 1976, were more or less copied from the Stockholm declaration. Article 13 stated that "the Polish People's Republic ensures the protection and rational development of the natural environment, constituting the nationwide wealth". Article 71 asserted that citizens "have the right to benefit from the values of the natural environment and the obligation to protect it" (Kabala, 1988: 6-7).

[66] Both Maurice Strong and Mustafa Tolba, the UNEP's managing directors in the 1970s, visited Poland several times during this decade. After one of his visits to Poland, Tolba told *Trybuna Ludu*, the central communist newspaper, that "Poland is one of the leading countries in the understanding and political realization of questions related to environmental protection" (quoted in Ochocki, 1985: 25).

[67] The most important of these reports was made by W. Brzeziński, S. Kozłowski, and Z. Wierzbicki, scientists affiliated with the Polish Academy of Sciences, at the beginning of 1980 (*Przyroda Polska*, 1981).

Gospodarki Terenowej i Ochrony Środowiska or MGTiOŚ).[68] It was reorganized in 1975 after an administrative reform and was henceforth called the Ministry for Administration, Territorial Management, and Environmental Protection (*Ministerstwo Administracji, Gospodarki Terenowej i Ochrony Środowiska* or MAGTiOŚ) (Alberski, 1996). Environmental protection interests thus achieved the formal recognition that the government should deal with this issue.

The environmental programme until 1990

In 1973 the elaboration of a long-term environmental protection programme began. The decision to commence this project was related to Poland's will to follow up the UN environmental conference in Stockholm in 1972 with a document outlining the national environmental protection strategy. Some 600 experts, mainly representatives of various scientific institutes, were engaged for the task and contributed more than 300 studies. In 1975 MAGTiOŚ synthesized the studies into a final version of the programme that came to include the following topics: the role of the socialist state for environmental protection, the use of policy instruments, organization and management, localization policy, science and technology, development of personnel, public involvement, and international cooperation. The programme also contained an assessment of the environmental situation in Poland and a list of proposed investment projects (Ochocki, 1985). I highlight some of the key issues of the document below.

Initially, the programme emphasized that the "socialist countries are facing the task of practically developing the Marxist science about the relation between society and nature" (MGTiOŚ, 1973: 9). The socialist system was said to have a greater potential for protecting the environment than the capitalist system, in which exploitation of natural resources has been significant. Only the socialist system, it was stated, permits a harmonious relation between man and nature. The Soviet Union was referred to as the model country, since it has "made the rational use of natural resources as an integrated part of the programme for building communism" (*ibid.*).

All central state organizations and ministries should be engaged in the realization of the programme because "the environmental question should be treated as one of the main directions of the state's activity". One of the urgent tasks for the central authorities was to develop the facilities to manufacture equipment for environmental protection (e.g. filters).

The state authorities should promote efforts by voluntary organizations to protect the environment. NGOs can perform important tasks such as popularizing environmental issues, conducting control activities, and arranging meetings, seminars, and competitions. Direct initiatives to improve the environmental situation, like planting trees, should be supported by the state authorities.

The document underlined that socioeconomic development has to be perceived from a new perspective. It was noticed that the social costs of economic activity, one such cost being environmental degradation, were not reflected in the firms' costs or in the national accounts. However, the document contained no precise recommendations on what should be done to improve the situation. The policy-makers argued that environmental fees should have an incentive effect. In other words, it ought to be cheaper for the polluters to control pollution than to pay the fee (*ibid.*: 42).

Environmental expenditure was to be increased, bringing it in line with the levels

---

[68] Simultaneously, a corresponding committee was set up within the parliament (Alberski, 1996).

budgeted in Western countries. The goal was devote 1.5 per cent of GDP to environmental protection (*ibid.*: 71). Almost half of the investment costs were to be covered by the state budget. It was forecast that the largest share of the investments until 1990 should be devoted to water management and water protection.

Environmental legislation, planning, and statistics should be developed "as has been done in other parts of the world" (*ibid.*: 14). It is interesting that, at the same time as the document referred to the Soviet Union as the model state for environmental protection, positive references to developments in the West were made. Besides the general statement saying that "organizations controlling the environment should be developed" (*ibid.*: 12), there was no mention about enforcement mechanisms.

The Statute on the Protection and Shaping of the Environment

The elaboration of a comprehensive environmental law started after the PZPR's seventh congress in December 1975. Of particular importance was the following resolution at this conference:

> We must give more attention than before to the protection and shaping of the environment. With this in view we must build our towns and villages and protect the aesthetic values of the countryside. This should be an important part of development planning, investment programming, and technology preference. We should take into account the importance of the problem; there should be prepared, with the help of scientists, a bill regulating environmental protection.[69]

The work started. Initially it was not clear what kind of bill was going to be drafted. Two concepts were put forward: a code and a general framework act. The latter would entail more or less copying the US National Environmental Policy Act, whereby more specific legislation would be added later. Ultimately, both of these ideas were rejected. According to Jerzy Sommer, who was a member of the team preparing the law, the code concept fell because the policy-makers thought that there was not sufficient experience in Poland to elaborate such a law. The concept of a general policy act of the American type was rejected because the policy-makers were convinced that the new act had to include a critical mass of regulations in order not to become empty words.[70] Instead, the policy-makers chose a third concept, a hybrid version of the two concepts, which combined general formulations with detailed regulation concerning protection of the environmental elements, policy instruments, etc.[71]

At the beginning of 1980 the parliament enacted the new law, "the Polish Party-state's single greatest effort to protect the environment" (Cole, 1995a: 338), which was given the name Statute on the Protection and Shaping of the Environment (*Ustawa o Ochronie i Kształtowaniu Środowiska* or UOKŚ). UOKŚ was a comprehensive law covering the entire field of environmental protection. All in all, there were eight titles with 118 articles. The most important parts of the document are presented below.

---

[69] VII Congress of the PZPR: basic materials and documents (quoted in Cole, 1995a: 322).

[70] Personal communication with Jerzy Sommer, 1995.

[71] Personal communication with Jerzy Sommer, 1995.

A. Environment and planning. A central feature of the Statute was that it called for an integration of environmental concerns into the socioeconomic plans:

Art. 5. 1: Environmental protection constitutes an essential element of national socioeconomic policy. Matters pertaining to environmental policy shall be included in the national socioeconomic plans, land-use plans, and normative statute, and will be taken into account in the activities of national organs, national economic units, and social organizations.
Art. 5. 3: The national socioeconomic plans shall take into account, as an integral part of the planning provisions, tasks, and means to ensure effective environmental protection and the effective elimination of activities with negative environmental impacts.[72]

B. *Protection of particular environmental elements.* There were nine chapters in the Statute dedicated to specific goals of environmental protection. These chapters provided a framework for further, more detailed, regulation. The following environmental elements were to be protected: the earth's surface and minerals, waters and marine environment, the atmosphere, flora and fauna, landscape values, and green areas. The statute also included chapters about protection against noises, wastes,[73] and radiation.

C. *Environmental duties.* The 1980 UOKŚ established the principle that all central, regional, and local governmental agencies, enterprises, and individuals have a duty to protect the environment. They were to do so by:
1. carefully siting production facilities to create the least environmental impact;
2. taking protective measures during economic activities;
3. restoring environmental conditions damaged by economic activity;
4. making use of new technologies to reduce environmental impacts of economic activities, especially waste-reduction, and waste-prevention technologies;
5. constructing, installing, and maintaining appropriate environmental protection equipment;
6. installing monitoring equipment and conducting necessary measurements;
7. complying with environmental protection requirements in planning, designing, and manufacturing machinery, equipment, etc.;
8. recycling wastes and effluents or ensuring their effective neutralization or disposal;
9. making use of scientific and technical progress and legal and administrative means of environmental protection.[74]

D. *Policy instruments.* The 1980 law reinforced the system of individual permits that had been introduced in the 1960s. It stated that the environmental departments in the 49 voivodships should issue permits that impose emission standards for a particular source of pollution. This permitting process should be guided by national environmental quality standards for water, air, noise, radiation, and vibration (Sommer, 1993: 4). The law obliged voivodship authorities to close industrial plants

[72] UOKŚ 1980, quoted in Cole, 1995a: 325.

[73] For the first time in Poland's history, waste management appeared in a legislative document.

[74] UOKŚ, Article 66.

when their pollution levels posed a direct threat to public health.

The Statute obliged economic entities to monitor their emissions of air pollution and to take measures to improve air quality:

> (...) the Council of Ministers was authorized to promulgate regulations establishing permissible concentrations of air pollutants and guidelines for measuring and monitoring pollutant levels in the atmosphere. The regional authorities were also given a substantial role to play in the protection of the atmosphere. Article 30 authorized them to regulate categories and levels of air pollutants (...). In the event of a violation of air pollution norms (...), the regional organ of state administration could indefinitely suspend the activities causing the violation until levels of air pollution were brought within standards (Cole, 1995a: 330).

The role of economic instruments was substantially reinforced. Existing fees for water use and pollution were extended with fees for disposal of waste and activities causing air pollution. Fees were to be paid by all economic entities for their use of natural resources. Also the fine system became much more comprehensive. Facilities that violated environmental conditions more were subject to fines not only for effluent discharges but also for air pollution emissions, noise-producing activities, chemical uses, and waste dumping activities. Collected fines and fees were earmarked for a new Environmental Protection Fund that would finance various environmental protection projects. This system of fees and fines became, together with the permit system, the primary tool of environmental protection during the 1980s (Cole, 1995a: 336-337).

According to the liability provisions in the Statute, a person or an economic entity that had caused damage due to pollution was mandated to pay costs connected with the elimination of the threat and the restoration of the environment to its pre-existing conditions.[75]

The 1980 UOKŚ reinforced the sanctions towards polluters that had been provided under the Penal Code of 1969[76] and Petty Offences Code of 1971. Three years imprisonment was in store for those who caused "potential danger", neglected their pollution control equipment, imported waste from abroad, or failed to sufficiently protect agricultural and forestry lands (*ibid.*: 335-336).

E. *Administration.* The Statute called for the creation of a State Environmental Protection Inspectorate (*Państwowa Inspekcja Ochrony Środowiska* or PIOŚ). It was to be responsible for (1) supervision of compliance with environmental legislation; (2) environmental monitoring; and (3) initiation of activities for environmental protection. The creation of PIOŚ was an embryonic enforcement mechanism in Poland's environmental protection system. However, it would definitely be premature to call PIOŚ an enforcement agency. For instance, it lacked the authority to levy fines and close down polluters. Furthermore, the Ministry of Administration, Territorial Management, and Environmental Protection as well as the Council of Ministers had the possibility to override decisions made by PIOŚ (*ibid.*: 337).

F. *Public participation.* The 1980 UOKŚ was a breakthrough, at least on paper, in allowing voluntary organizations and private citizens to participate in the

---

[75] UOKŚ, Articles 80 and 82.

[76] Under this Code, polluters could be imprisoned for up to ten years for intentionally posing a threat to human life, health, and property.

environmental policy process. Non-governmental organizations "were empowered to file lawsuits to suspend environmentally threatening economic activities and order restoration" (*ibid.*: 338). Moreover, these organizations had to be informed and their comments considered before the authorities could initiate activities potentially harmful to the environment.

In sum, the 1980 Statute came about as a result of the communist party's will to promote environmental protection in the mid-1970s. The most important feature of the law was that it introduced a comprehensive system of environmental fees, fines, and funds.

Other initiatives

Several other environmental policy initiatives in the 1970s deserve to be mentioned. In 1970 the government created the Polish Committee for the Protection of the Human Environment to coordinate and promote environmental policy initiatives within the government (Alberski, 1995).[77] In the same year the concept of environmental protection (*ochrona środowiska)* was used in Poland for the first time. It appeared in a resolution from the Polish Academy of Sciences (*Polska Akademia Nauk* or PAN) in which the committee "Man and Environment" was appointed (Cole, 1995a: 319). The first Polish environmental magazine, *Aura*, a monthly, was established in 1972. It was a comparatively small magazine compared to the main nature protection magazine, Polish Nature (*Przyroda Polska*), issued by LOP since 1957 (Ochocki, 1985: 101). The main technical magazine, Technical Survey *(Przegląd Techniczny*), became an increasingly important forum for information and debate on environmental issues. A department to deal with environmental issues (Department for Perspective Spatial Planning or *Zespół Planowania Perspektywnego*) was set up in 1972 within the State Planning Commission.[78] The socioeconomic five year plan for the period 1971-75 included for the first time a chapter on environmental protection, mainly focused at water management, water protection, and dust reduction (Alberski, 1996: 112). In 1972 the Main Statistical Office (*Główny Urząd Statystyczny* or GUS) started to produce environmental statistics for "internal use". Moreover, a large number of books on environmental protection were published.[79] More than 7,000 publications about environmental issues were published between 1945 and 1974 (Chmielewski, 1988: 51). As a rule, information about the environment "was not secret but it was not open to the public".[80]

In 1974 two important statutes were enacted: the Building Law and the Water Law. The former contained provisions about the necessity to design, use, and construct buildings in such a way that environmental quality was ensured. The latter had a relatively comprehensive approach to water management. Among other things, it

---

[77] The committee's main task was to prepare for Poland's participation in the Stockholm conference in 1972. It was dissolved in 1972 (personal communication with Robert Alberski, 1995).

[78] Personal communication with Andrzej Kassenberg, 1996.

[79] Characteristic of Polish environmental literature was an inclination to evade national problems and to illustrate the ecological crisis abroad (Delorme, 1981). There was more information about smog in London than air pollution in Upper Silesia.

[80] Personal communication with Andrzej Kassenberg, 1995.

required protected zones around water intakes and it introduced fees for consumption of water and disposal of waste water. These fees were to feed a Water Management Fund from which money for financing water quality projects could be derived (Cole, 1995a: 321).[81] In 1976 the Environmental Impact Assessment (EIA) concept was introduced as part of the land-use planning law (World Bank, 1992b: 27). Each voivodship had by 1975 established a Department for Environment Protection and Water Management (*Wydział Ochrony Środowiska i Gospodarki Wodnej* or WOŚiGW) together with their Centres for Measurement and Control of the Environment (*Ośrodek Badań i Kontroli Środowiska* or OBiKŚ) (MGTiOŚ, 1975: 4). Most of the heavy polluters had established their own environmental protection departments by the late 1970s (Górka and Poskrobko, 1987: 99).

3.7. Implementation

The 1970s were a setback in terms of realization of the ambitious plans. Poland's environmental protection actions lagged behind its rhetorical flourishes; environmental policy was characterized by an implementation deficit.[82] Consequently, the envisaged improvement of the environmental situation failed to materialize and the environmental situation continued to deteriorate. Between 1975 and 1980, emissions of gas pollution increased from 3 million tonnes to 5.1 million tonnes. Also dust emissions went up during this period (Alberski, 1996: 62). Likewise, there were no signs of improvement in surface water quality. One example of this is that the first beaches on the Baltic Sea coast had to be closed in 1977 (Chmielewski, 1988: 64).

The reasons for this failure are manifold. First of all, the most important decisions regarding the economic strategy, investments, and the siting of large industrial objects (such as power plants, steel mills, and mines) were taken above the ministry in charge of the environment (Alberski, 1996: 100). The environmental administration had no impact on those policies that caused the environmental damage (*ibid.*: 110). The Ministry for Administration, Territorial Management, and Environmental Protection mainly dealt with "territorial management" (which corresponds to local development), not environmental protection. The section dealing with the environment was weak and understaffed; only some 15-20 persons worked with this issue.[83] Moreover, many important issues regarding natural resources management (forestry, nature protection, water management) remained outside the Ministry's jurisdiction. The creation of

[81] The introduction of economic instruments was inspired by developments within the economic sciences in the Soviet Union in the 1960s and 1970s. Research on the use of economic instruments in environmental protection was conducted by Fyedorenko, Gofman, Gusev, Zamin, and others. Furthermore, a department for ecological economy was created at the Central Mathematical and Economic Institute of the Academy of Sciences (Fyedorenko, 1976). The work of the economists included general models for improved environmental performance of the economy, the inclusion of the environment into economic calculations, definition of economic losses due to environmental degradation, economic assessment of natural resources, the improvement of the planning process, and increased economic incentives for the enterprises to reduce their negative environmental impact (Stoklasa, 1980).

[82] In Weale's (1992: 17) definition, "implementation deficit arises when legislative and policy intent is not translated into practice". Implementation deficit was also a common feature of many western countries, for example, the USA and Germany.

[83] Personal communication with Janusz Żurek, 1995.

MGTiOŚ was preceded by the abolition of the CUGW. There is enough evidence to conclude that most environmental policy-makers in Poland perceived the administrative reorganization in 1972 as a step backwards rather than forwards.

Energy prices were kept artificially low. Consequently, the energy intensity of the economy continued to increase.[84] Industry did not comply with the environmental legislation, and there were virtually no enforcement mechanisms at hand. Siting of new industry perpetually breached the environmental regulations. The most notorious example is the gigantic steel plant Huta Katowice, which was located in the industrial centre in Upper Silesia, the most polluted region, in 1973. The entire project was constructed outside existing land-use plans as well as the regional and national socioeconomic plans (Cole, 1998: 97). Planned environmental investments and modernization projects were postponed due to the deteriorating economic situation in Poland. Environmental expenditures decreased dramatically compared to the foregoing five-year period. According to Górka (1981), environmental expenditure as a share of GDP decreased from a level close to 1 per cent between 1971 and 1975 to about 0.4 per cent between 1976 and 1979. Another phenomenon appeared in that the budget for environmental investments was not fully spent. For instance, only half of the money budgeted for construction of industrial and municipal waste water treatment plants was used in this period (Aura, 1980c).

The foreign credits that Poland received throughout the 1970s could have given impetus for the modernization of the economy in an environmentally friendly way. But the funds were spent without any consideration for the environmental consequences of the investments. Nobody made environmental protection a condition for the credits. Hence, both the Polish government and the western lenders were responsible for these neglects.[85]

3.8. The dominating belief system

The actors within the nascent environmental policy subsystem in the 1970s did not have the properties of an advocacy coalition. Rather it should be seen as a loosely organized issue network with a belief system that can be reconstructed in the following way.

Deep core beliefs

Socialism was seen as infallible, beyond empirical challenge. It was believed to offer a fundamental solution to all problems, including environmental ones. In Kornai's words:

> Official ideology suggests that the creation and maintenance of the socialist system is in itself a thing of value, a primary good. Even if it fails at a specific point to yield the performance of which its innate properties render it capable, it will yield it sooner or later. After all, any single factor of performance (e.g. material welfare, efficiency, or fair distribution) is only instrumental in nature and exists to further the real purpose of creating and defending the socialist order. These factors will certainly appear as a by-product of the socialist system,

---

[84] In contrast to the western world, Poland was not affected by the oil price shock in 1973.

[85] Personal communication with Waldemar Michna, 1995.

even if there is a delay before they do so. Of greatest importance is the simple fact that socialism has been won (Kornai, 1992: 52-53).

Hence, socialism *per se* was perceived as the solution to all policy problems. All problems were structured according to the socialist formula:

The ideas and methods termed "scientific socialism" in the official ideology ensure intellectual superiority to those who know and employ them over the exponents of any other ideas and methods, since it supplies them with a reliable compass for comprehending any new situation and identifying the new tasks it poses (Kornai, 1992: 56-57).

"Scientific socialism" – as it was interpreted in the Soviet bloc – was a petrified policy theory that did not allow unstructured problems to exist. When unstructured problems appeared they were treated as structured or simply denied (see Section 2.2).

Policy core beliefs

The proper relationship between environment and economy. Environmental policy was perceived as a policy area within its own right. The main task for environmental policy was seen as promoting the development of water management, the construction of sewage treatment facilities, and the protection of nature. It was believed that these tasks could be carried out in isolation from general economic development. No attention was given to the need to integrate environmental concerns into sectoral policies.

The belief was widely held that Poland first should become rich and then take care of its environment. Economic growth was seen as a precondition for environmental protection. In the 1970s the idea prevailed that the most developed nations in Europe did not start to take environmental protection seriously until they had become rich. No country had ever given sufficient attention to environmental protection in the early phase of industrialization. Therefore Poland should first catch up with the West, then invest in the environment. Virtually nobody questioned this assumption.[86] Not surprisingly, "Limits to Growth" by the Club of Rome, which appeared in Polish in 1973 (Tobera, 1988: 69) received a hostile response in Poland. A number of professors, nature conservationists, politicians, and others referred to it as "neo-colonialism".[87]

Whose welfare counts? According to the prevailing ideology it was the welfare of the workers that was the first priority for the state.

The seriousness of environmental problems. There was a general recognition that the environmental situation was bad in several respects.

Basic cause of environmental problems. Flaws in implementation were to be corrected by "more of the same" rather than by conceptual changes. The main explanation advanced in the discussion about the causes of the implementation deficit in the latter part of the 1970s can be summarized as follows: lack of money and lack of specialized enterprises producing the required equipment. In addition, individual participants in the policy process were often blamed: the ministry miscalculated the investment plan, the factory management did not use the filters properly, efforts by

[86] Personal communication with Waldemar Michna, 1995.

[87] Personal communication with Tomasz Żylicz, 1995.

the regional authorities were not rigorous enough, etc.

Proper scope of governmental vs. market activity. This issue was not addressed in the environmental policy discourse in the 1970s.

Proper distribution of authority. It was assumed that the central authorities, especially the ministry in charge of the environment, should play the most important role in the environmental policy-making process.

The choice of policy instruments. There was wide support for the use of national environmental quality standards and the permit system. The interest in economic instruments increased in the 1970s, which is reflected by the introduction of water fees in the middle of the decade.

Method of financing. State subsidies were widely acknowledged as the most appropriate method of financing environmental protection investments.

Ability of society to solve the problems. The application of end-of-pipe technology was seen as the most appropriate way of achieving emission reductions. A frequently cited slogan was that technology should repair what technology had damaged.[88] In other words, there was a general atmosphere of technological optimism.[89]

The environmental policy actors were convinced that the existing political and economic system had an untapped potential with respect to environmental protection.[90] Even those who were critical of the environmental performance shared the optimistic convictions of the ruling elite. There was general consensus that the situation could be improved within the framework of the existing system because there was an unused potential waiting to be unleashed. In particular, there were very high expectations that the new comprehensive environmental law – which was under preparation during the latter part of the 1970s – would bring about necessary improvements.

Who should participate? It was generally assumed that experts of various kinds should play the most important role in the policy-making process. The general public and voluntary organizations were invited to participate if that was required for the implementation of environmental policy. Thus, their input in the agenda-setting phase and policy adoption phase was not seen as important.

Secondary aspects beliefs

Institutional aspects. Political loyalty was more important than professional knowledge. A secretary of the PZPR, based in Warszawa, once said: "The party's personnel policy is: No one has to be competent in any post; he merely has to be loyal" (quoted in Kornai, 1992).

Priority problems. In the 1970s, water management and nature protection were seen as the most important tasks to be dealt with.

Policy-oriented learning

To a large extent, Poland's environmental policy "take-off" in the 1970s was an imitation of developments taking place simultaneously in the West. This process was

---

[88] Personal communication with Stefan Kozłowski, 1995.

[89] See, for example, Ochocki, 1985.

[90] Personal communication with Andrzej Kassenberg, 1995.

made possible by the détente between the superpowers in this decade. The Polish leadership learned that the promotion of environmental issues could be used as a strategy for increasing the political legitimacy of the socialist Polish state abroad.

3.9. Summary and conclusions about policy change

This chapter began by identifying relatively stable parameters in Poland and the main external events in the period 1945-80. It then concluded that the main features of environmental protection before 1970 were nature conservation and water management. The promotion of environmental policy in the early 1970s included the elevation of environmental policy to ministerial level (1972), the adoption of a long-term environmental programme (1975), and the preparation of a comprehensive environmental framework law. However, Poland's environmental performance throughout the decade did not improve, rather the opposite. Environmental protection was subordinated to the goal of industrial production. The environmental discourse was characterized by the ideological assumption that socialism is more environmentally benign than capitalism. There was also a considerable degree of technological optimism. It was assumed that the central authorities should play the most important role in environmental protection and that the most proper way of financing investments was state subsidies. Environmental policy was determined by a narrow group of experts in isolation of the general public and industry.

The main driving forces for the environmental policy developments in Poland in the 1970s came from abroad. Poland was influenced to a large extent by the environmental upsurge that took place in western Europe and the USA in the late 1960s and early 1970s. The event that provided the greatest impetus for policy development in Poland was no doubt the United Nations Conference on the Human Environment in Stockholm in 1972. Poland took the preparations for this conference seriously, which is indicated by the fact that major policy initiatives in the early 1970s coincided with the preparations for this conference. Hence, to a large extent, the environmental agenda was imported. Environmental policy entered the domestic agenda via the foreign policy agenda.

4. NATIONAL ENVIRONMENTAL POLICY IN THE 1980s

This chapter addresses national environmental policy in the 1980s. Section 4.1 is an introduction to the 1980s, which contains the main external events of the period. Section 4.2 deals with the policy process from three perspectives: actors, issues, and interactions. Special attention is paid to the role of the Polish Ecological Club and the impact of Chernobyl. Section 4.3 presents the major policy documents of the 1980s. Section 4.4 focuses on implementation. Section 4.5 deals with the Environmental Protection Advocacy Coalition and its belief system and discusses policy-oriented learning. Section 4.6 draws conclusions about the main driving forces for policy change and summarizes the chapter.

4.1. Introduction

By the end of the 1970s, Poland's economic situation had deteriorated sharply. Public discontent with falling living standards resulted in a national wave of strikes in August 1980. For sixteen months in 1980-81, the independent labour union Solidarity (*Solidarność*), led by Lech Wałęsa, dominated the political scene in Poland. Solidarity had over nine million members, including one million communist party members (Millard, 1994: 15). Solidarity provided the impetus for two parallel processes: the self-organization of many spheres of society and the development of a new critical consciousness about the economic, political, and social problems of Polish society. Also environmental concern increased substantially, reflected by the creation of the Polish Ecological Club. Solidarity was abruptly removed from the political scene on the night of 13 December, 1981 when General Wojciech Jaruzelski imposed martial law and thereby restored the hegemonic role of the communist party (Gerrits, 1990). Martial law was abolished in the summer of 1983 but Solidarity remained an illegal organization for virtually the rest of the decade. An economic reform was launched by the authorities in 1982 in order to increase the autonomy of state enterprises (Balcerowicz, 1995: 292).

The final years of communist hegemony in Poland were characterized by political and economic relaxation. Gorbachev's perestroika in the Soviet Union allowed Poland to gradually extend political freedom. In 1987 the government arranged a referendum to obtain public approval of a macroeconomic stabilization programme. The people voted no, because of its distrust of the regime (Sachs, 1994: 36). The last socialist government, that of Mieczysław Rakowski, formed in October 1988, introduced substantial economic liberalization with respect to the private sector and foreign trade (Balcerowicz, 1995: 292). However, the economic situation deteriorated, the Polish people continued to look upon their rulers with deep mistrust, and the political conditions became more and more fragile. A new wave of strikes started in 1988. The government agreed to initiate a dialogue with Solidarity about the future economic and political development of the country. Preparations for Round-table discussions between the two parties started during the autumn of 1988. The Round-table was held between 6 February and 5 April 1989. Negotiations were held in a number of working groups on issues such as political reform, economic reform, trade union pluralism, education, health, etc. One working group was devoted to environmental policy reform (Millard, 1994: 61-62).

4.2. The policy process

4.2.1. New actors

The environmental policy network grew rapidly in the 1980s. Within the Polish Academy of Sciences, a committee for environmental engineering (*Komitet Inżynierii Środowiska*) was set up in 1982 (Aura, 1982b). The same year, the Main Technical Organization (*Naczelna Organizacja Techniczna* or NOT) established two new committees: one for environmental protection and one for water management (Aura, 1982a). In 1983 journalists specializing in environmental issues created a club called *Ekoś* to facilitate the exchange of information and ideas (Aura, 1983b).[91]

At the local level, a growing tendency towards organized efforts to address environmental problems was discernible. In 1982 two environmental organizations in Szczecin, the Club for Protection and Shaping of the Environment and the local branch of LOP, initiated legal action against Police chemical plant near Szczecin (Karczewski, 1984). The organizations did not demand closure of the company but called for radical measures to improve the environmental performance (particularly with respect to air and water pollution) and for economic compensation to people who had been exposed to harmful pollution. The litigation, a unique phenomenon in the socialist bloc, became a long and arduous process, which would continue throughout the remainder of the decade.

The nuclear catastrophe in Chernobyl in 1986 resulted in a radicalization of the environmental movement and the creation of a number of new environmental organizations in the latter part of the 1980s. Among these were: Wielkopolska Ecological Seminarium (*Wielkopolski Seminarium Ekologiczny* or WSE), the Polish Ecological Party (*Polska Partia Ekologiczna* or PPE),[92] Freedom and Peace (*Wolność i Pokój* or WiP),[93] the Silesian Ecological Movement (*Śląski Ruch Ekologiczny* or ŚRE),[94] and the Green Federation (*Federacja Zielonych*).[95] The last organization was led by young people associated with the PKE who sought a new way of acting. Piotr Gliński, a sociologist specialized in the Polish environmental movement, explains: "One of the reasons why Green Federation was created was that the Polish Ecological Club had decided to be an expert and elite organization. The young people from the Green Federation wanted to be independent and more able to realize values which

---

[91] The club was officially associated with the League for Environmental Protection.

[92] Created in 1988 (Alberski, 1996).

[93] WiP was an oppositional peace movement dominated by students who carried out protests against, for example, the chemical plant Siechnice outside Wrocław and the nuclear power plant construction in Żarnowiec (personal communication with Radosław Gawlik, 1995).

[94] ŚRE was based in Katowice and organized open manifestations against the environmental degradation of that region.

[95] Created in 1988.

were typical for the new social movements in the West".[96] At the end of the 1980s there were more than one hundred organizations and groups, most of them independent, actively involved in environmental protection.

However, the most important new actors in the 1980s were Solidarity and the Polish Ecological Club, two actors who will be described in some detail below.

Solidarity and the environment

Environmental protection was, from the very beginning, an integrated part of Solidarity's general strategy for a renewal of Polish society. In the Gdańsk agreement[97] the government promised Solidarity to guarantee "clean air, soil, and water – particularly in the coastal areas" (Polityka, 1980). The importance of this brief passage was that it placed environmental protection on the agenda of change.

On 28 July 1981 a national environmental commission entitled Man and the Environment was set up at the Social and Union Research Center (*Ośrodek Prac Społeczno-Zawodowych* – or OPSZ) by Solidarity's National Coordination Commission (*Krajowa Komisja Porozumiewawcza* or KKP) (Serafin, 1982: 176). Two issues were considered to be of particular interest: the link between environmental protection and economic revitalization and the public health consequences of environmental degradation.

At Solidarity's first National Congress of Delegates in October 1981 two resolutions directly referred to the environment. Number 15 stated that "in the face of the biological threat to the nation, ensuring public health is of special concern to the Union". In resolution 16 Solidarity confirmed its general commitment to "struggle for an effective protection of man's environment" (*ibid.*: 176).

Within Solidarity there were those who thought that the union, which was present at virtually all places of work in Poland, had the potential to become a mass movement for the improvement of the environmental situation. In a letter published in *Aura*, Piotr Tobera (1981), a researcher at the Sociological Institute of Łódź University, suggested that Solidarity's regional structure and high prestige constituted a chance for the greening of the whole state. A public appeal made by Solidarity's environmental commission reveals that it had the ambition to mobilize the whole union in the cause of environmental protection:

Remember! At all work places, during play and leisure time, in the forests, at the lake, the park, at sea – everywhere your contribution to the great work of protecting the environment is needed. Join a club, a commission, or a section for environmental protection if something like that exists at your work or in your region. If there is not, try to organize something. Look for people of good will around you, in your neighbourhood, who would like to do something for the benefit of the public health and the beauty of the nation (Tobera, 1981).[98]

---

[96] Personal communication, 1995.

[97] In this agreement of August 1980, the government acceded to all of Solidarity's twenty-one demands. Most importantly, the agreement accepted the principle that independent organizations, such as labour unions, might exist (Millard, 1994: 14).

[98] In addition, the environmental commission declared that it was open for cooperation with other organizations, partly or fully dedicated to environmental protection, such as the Polish Ecological Club, the Polish Tourist Organization (PTTK), the environmental committees of NOT, and the League for Nature Protection (LOP) (Krajowa Komisja, 1981).

By and large, the vision of Solidarity as a mass organization for environmental improvement failed to materialize. Solidarity was not transformed into a green movement, and environmental protection never became a priority issue for the union.[99] Still, Solidarity had a huge impact on environmental policy development in 1980-81. Firstly, environmental protection was placed for the first time in a larger context. Secondly, Solidarity temporarily abolished PZPR's information monopoly. Solidarity partly succeeded in abolishing censorship. Access to all sorts of data increased. The dimension of the environmental degradation in the country had been revealed once and for all. One of the main fora where this debate occurred was the monthly environmental journal *Aura* in which open letters to the authorities from different organizations and proposals from various policy-makers and independent researchers were regularly published. In a leader in mid-1981 (*Aura*, 1981a), *Aura* revealed its earlier problems with censorship:

> Before, we could not write about health in concrete terms, we could not point to specific factories or to people whom were affected. The Office for Press Control obliged the censors to eliminate everything that connected health problems with a defined economic activity.[100]

Furthermore, more and more international literature and reports on various environmental issues appeared in Polish translation. For example, more or less all reports published by the Club of Rome were translated into Polish. In 1986 the first State of the World report by the World Watch Institute was available in Polish (Żylicz, 1989).

Thirdly, Solidarity created political space for independent environmental initiatives.

The Polish Ecological Club

In an interview in *Przegląd Techniczny* (Lipszyc, 1980) in the autumn of 1980, Michalina Białecka and Krystian Waksmundzki, two of the founders of the recently established Polish Ecological Club, explained why Poland needed an independent environmental movement. Firstly, the ineffectiveness of earlier activities. During the 1970s, many environmentally concerned members of the scientific community became frustrated because of the ineffectiveness of their efforts. "No matter which methods we used – open letters, memorials, scientific reports, petitions, meetings with the authorities, work within the municipal councils, and various scientific councils – we achieved no results. For how long could we carry on like that?" Secondly, the lack of an independent environmental NGO. According to Białecka and Waksmundzki, Poland was suffering from the absence of a strong, courageous, and independent environmental organization prepared to challenge official policy. Existing organizations, like LOP, had failed to fulfil these requirements. Thirdly, the lack of reliable data. Not only had the authorities blocked information, but certain

---

[99] It is difficult to find any statements where the leading personalities within the union referred to the environment.

[100] The leader also mentioned some of the articles which had been censored before August 1980: the connection between chemical pollution and cancer at the plant "Boruta" in Zgierz, health effects of leaking radioactivity from the storage of nuclear waste in a uranium mine in the Sudeten Mountains, and the health situation of the workers at the copper smelter in Legnica (*Aura*, 1981a).

environmental information had been falsified and the policy-makers had not always been provided with accurate data. This problem could be mitigated by an honest and open criticism from an independent organization. Fourthly, the isolated position of environmental policy. The environmental policy of the 1970s was, in Białecka's and Waksmundzki's view, formulated and implemented in isolation from the general development of the country. A solution to Poland's environmental problems would require the inclusion of social, economic, legal, and above all, political dimensions in the policy process.

The birth of the PKE, the party-state's first independent environmental organization, dates back to a meeting held in Kraków, 23 September, 1980, which gathered together several hundred representatives of the local scientific, cultural, and environmental circles. The central ideas of the new club were formulated in an open letter addressed to the Sejm. It made clear that it is a basic human right to live in an environment "not harmful to public health and not threatening flora and fauna, the landscape, historical monuments, and culture", and that environmental degradation in Poland posed a threat on such a scale that the biological and cultural existence of the Polish nation was in danger. This situation arises from "[t]he centralized and bureaucratized administration, the indifference of the political authorities *vis-à-vis* scientific warnings about the ecological destruction of the country, carelessness of the economic authorities, disinformation to the public, and arbitrary decisions from above (...)". Existing authorities have failed because they have been "unable to execute the law, act independently, and effectively fulfil their obligations". Furthermore, they have been deprived of the right to inform the general public about the shortcomings (Maciejewski, 1980).

The first large-scale campaign launched by PKE was directed towards the aluminium smelter Skawina, a notorious polluter situated in the outskirts of Kraków. The plant was erected in the early 1950s and had been based on technology outdated from the outset. In the mid-1970s the dramatic environmental effects of aluminium production were revealed: the level of fluorine in the soil in the local environment was 5 to 35 times higher than the Polish average, and half of the cattle grazing in the surroundings had fluorosis. Moreover, the employees of the plant were suffering from various pollution-related diseases. In 1976 the local authorities of Kraków obliged Skawina to modernize the electrolysis unit in order to achieve a substantial reduction in fluorine emissions by 1980. However, virtually no action was taken and fluorine emissions remained unabated (Bieroń and Garścia, 1981).

The campaign against Skawina aimed at an immediate closure of the plant. PKE realized that it could not achieve this goal on its own. Environmental specialists from the Polish Academy of Sciences, the Academy for Mining and Metallurgy, and Solidarity were asked to collect reliable information about the environmental situation, and close cooperation with the local press and other mass media was established to mobilize local opinion against the plant.

In November 1980, some people living in the vicinity of the plant brought a court action against the management of the smelter as well as the local authority in Skawina, demanding compensation for health damage caused by toxic waste from the mill (Economist, 1981). The campaign gained even more momentum in December the same year when the local authorities created an environmental council which made clear that Skawina was the most urgent environmental problem for Kraków. Within the council, an independent expert team, comprising representatives of science and industry, rapidly came to the conclusion that the plant should have been closed already in the early 1970s. In the beginning of January 1981, the president of Kraków decided to stop aluminium production using the right to close excessive polluters afforded by the newly adopted Statute on the Protection and Shaping of Environment (Bieroń and Garścia, 1981).

Skawina had a catalytic impact on the environmental movement. The campaign had clearly shown that public pressure mattered and that real environmental improvements were attainable. Furthermore, it inspired many local groups to take action against local environmental degradation.[101]

Environmental issues went through a rapid issue-attention cycle in 1980-81. The following stages can be identified. In the autumn of 1980 there was a period of alarmed discovery and euphoric enthusiasm. In the first half of 1981 this euphoria was gradually replaced by an insight that further environmental improvements would have to be followed by certain economic sacrifices. For example, the decision to close the aluminium smelter in Skawina imposed further strains on the Polish economy. According to the Ministry of Metallurgy, the annual cost for replacing the lost production (Skawina had half of Poland's aluminium production capacity before the closure) with imported aluminium would amount to $50 million. The economic complications of the closure was commented on by a leader in *Aura* in the following way:

In the present situation, when there is a struggle for bread, energy, and raw materials - in short, the society's basic needs - environmental measures will have to give way to addressing the most urgent problems. The closure of the aluminium plant in Skawina, although commonly acknowledged as a success for the environmental movement, cannot be a testimony that we can afford to immediately eliminate all the negative consequences brought about by industrial civilization (Domański, 1981b).

Then ensued a period characterized by a gradual decline of intense public interest. *Aura*'s editorials in the autumn of 1981 aptly reflect the prevailing mood of Polish society. The October issue made clear that the environment was no longer on the agenda:

Struggle for bread, energy difficulties, falling efficiency of work and production - environment is not on the agenda. Lack of light. Lack of basic products. Our efforts are focused on increasing production wherever possible. This is the main reason why nothing is happening within environmental protection (Domański, 1981c).

The dismal situation was confirmed by the editorial of the November issue. It concluded that neither PZPR's IX extraordinary congress, nor Solidarity's first national congress had focused sufficient attention to the environment. The leader concluded: "In times when each citizen has the right to only one bar of toilet soap every two months, environmental protection is more and more pushed into the distant future" (Aura, 1981c).

In many respects it is fair to compare PKE to the western environmental movement of the early 1980s. However, in two respects the club was different. First, it had a professional standing. In Kabala's words, "PKE members sit on city councils, work on the National Planning Commission, belong to the Polish Academy of Sciences and hold university posts in almost every discipline, from geology and metallurgy to

---

[101] For example, in Tymbork in southern Poland, the local PKE group organized a campaign against a bitumen factory polluting a local river used as a water source for a downstream fruit juice factory. The plant was ultimately forced to find a new location (Serafin, 1982: 183).

sociology and architecture" (Kabala, 1988: 5-6). Second, it explicitly wished to remain a small organization, dominated by experts, rather than becoming a mass movement.

It is evident that the first priority for PKE (as well as for the rest of the independent movement) was public health. Bearing in mind the low quality of drinking water and the serious air pollution problems in most urban areas, this was a natural choice. The link between health and environment also had a strategic dimension since it offered an opportunity for the environmentalists to open up a dialogue with Solidarity for whom the health of the workers was a central concern.

Another element of PKE's environmental ideology was the protection of the cultural heritage. Of particular importance was Kraków, "a sanctuary of the national identity", where the historical monuments were being destroyed by air pollution. PKE was also concerned about the destruction of the historically shaped landscape (Aura, 1981b).

PKE claimed that environmental protection was a "patriotic obligation". PKE's linkage of environmental protection and patriotism in its action rhetoric may have been part of a strategy to expand the issue to a larger group. As Cobb and Elder (1983: 156) have noted, "an important strategy in issue expansion is to associate a particular issue with emotionally laden symbols which have a wide acceptance in society".

The experiences, which were gained within the environmental movement during 1980-81 had important implications for its work during the rest of the decade. As regards *policy*, it seemed clear that the system had a potential for increased environmental performance, something that had been illustrated by the closure of Skawina, and PKE's victory in Tymbork (see footnote 101).

The second aspect was *political*. Some people in the opposition had discovered that the environment movement was an excellent platform in the political struggle for freedom and democracy.[102] Worrying facts about environmental destruction could be used to discredit the socialist system and to mobilize people to participate in the struggle for the abolishment of the system.[103] Small actions could have wide political repercussions, even outside Poland. As Grzegorz Peszko, a member of the Green Federation in the late 1980s, concludes: "Ten people standing on the market square was enough to attract the attention of the whole world".[104]

Hence, three different groups had appeared within the environmental movement, groups with different aims and strategies: (1) the policy-oriented environmentalists, who focused on incremental changes within existing structures; (2) members of the opposition, which used the environmental movement as a platform for the struggle *against* the system; and (3) local environmentalists, who were primarily concerned with eliminating direct threats posed by industrial pollution.

---

[102] Personal communication with Andrzej Kassenberg, 1995.

[103] Often the politically oriented environmentalists exaggerated the scale of environmental destruction. One obvious exaggeration was Krystian Waksmundzki's statement in *Przegląd Techniczny* (Lipszyc, 1980) that the annual gas emissions (sulphur dioxide, nitrogen oxides etc.) in Kraków amounted to 31 million tonnes, which in fact corresponded to more than five times Poland's total gas emissions. On another occasion, PKE claimed that 40 per cent of Poland's territory was turning into steppe (Lewiński, 1981).

[104] Personal communication, 1995.

4.2.2. New issues

New issues emerged on the environmental agenda in the 1980s. The most important among these were (1) the integration of environmental concerns into economic policy-making and (2) nuclear power.

Environment and economy

Shortly after the economic reforms had been launched in 1982, a lively debate ensued about the environmental consequences. The debate centred round the following question: would the increased financial incentives for the polluters improve or worsen their environmental performance? The general conclusion was that economic reform would, on the one hand, increase the effectiveness of economic policy instruments like fees and fines; on the other hand, a higher degree of cost awareness would provide the polluters with incentives to save on environmental expenditures.

The rate of the environmental charges was increased three-fold in 1982 (Aura, 1983a).[105] Moreover, in 1983 it was decided that fines incurred for pollution must be paid from company *profits* (Kabala, 1988: 4).[106]

The interest in identifying losses caused by environmental degradation increased in the 1980s. It was generally argued that the value of the economic losses was much greater than the value of environmental expenditure.[107] Hence, increasing environmental expenditure would be profitable in economic terms. Some of the estimates of the economic losses are presented below. The Cracow Academy of Economics estimated that the 1985 losses accounted for more than 25 per cent of national income. For the Kraków province the losses were equivalent to approximately 34 per cent of regional net production. Symonowicz's estimations pointed to losses corresponding to 7-9 per cent of the national income in 1983 (Familiec *et al.*, 1991). The economist Kazimierz Górka argued that the Katowice province annually lost up to 30 per cent of the agricultural production and half of the forest growth. Moreover, usage of buildings, instruments, and machines was 10-20 per cent lower than in the rest of the country, particularly due to corrosion caused by air pollution and the use of salinated water. Furthermore, the transportation means had a reduced life span (Górka, 1984). Professor Jan Juda of the Technical University in Warszawa argued that in the early 1980s, the yearly losses due to the sulphur dioxide emissions – in terms of public health, reduced agricultural production, forest damage, and corrosion – corresponded to $1 billion (*Aura*, 1982a: 9).[108] The different

---

[105] The following funds, directly or indirectly related to environmental protection, were in operation from 1982 (Radecki, 1983): the Fund for Water Management, the Fund for Environmental Protection, the Fund for Agricultural Soils, and the Forest Fund. The first three funds were divided into central and regional funds, the last one only operated on the central level.

[106] Thus, until 1982 it had been possible for the polluters to add the fines to their general costs.

[107] According to Górka (1985: 131), the economic losses were four times higher than expenditure on investment and maintenance related to environmental protection.

[108] The World Bank (1992b: 10) considers these estimates to be substantially too large. "In very roughs terms, it seems reasonable to conclude that the production cost approach to measuring the losses due to environmental damage will yield estimates that should not exceed 2.5-3 per cent of GDP, of which approximately 1-1.5 per cent of GDP is due to health effects of air pollution, and another 0.5-

outcomes of these calculations indicate that there was no clear idea how the estimates should be made. The need for adequate methodologies for correctly assessing the losses was widely acknowledged.

In the last few years before the demise of the socialist system the Ministry of Environment and Natural Resources argued that the establishment of companies specializing in pollution control technology is good business for the country. The argument was also advanced that Poland urgently needed foreign technology, for example, via joint ventures and international financial assistance, to deal with its pollution problems. Western proposals, most notably from the World Wide Fund for Nature (WWF), to reduce the Polish foreign debt according to the model debt-for-nature swap was met with enthusiasm (MOŚiZN, 1988a).

In connection with the celebrations of World Environment Day on 5 June 1985, the PKE organized a national scientific conference entitled "Eco-development[109] - a chance for survival of civilization" (*Ekorozwój szansą przetrwania cywilizacji*).[110]

A number of policy proposals were put forward at the conference. For example, it was suggested that 35-40 per cent of the territory of Poland be protected. Clean, non-waste technologies should be promoted rather than "half-measures" such as high chimneys and the dilution of waste water. The need for an interdisciplinary approach to environmental research was stressed. Spatial planning and scientific research should be integrated and "eco-regions" should constitute the fundamental unit for spatial and regional development.[111]

The conference is considered to be a milestone for the PKE. After the conference, the PKE began to take on all the appearances of a shadow ministry of environment. Furthermore, the conference contributed to an integration of the scientific community and strengthened the links between the PKE and the academic world.

The second national congress of the PKE, held in Kraków in 1987, approved so-called programme theses concerning eco-development. This coincided with the publication of the so-called Brundtland commission's report on environment and

---

0.8 per cent of GDP is due to the material costs associated with high levels of water salinity and biological oxygen demand (BOD) in the country's major rivers."

[109] Eco-development is the direct translation of the Polish word *ekorozwój*. A common translation is also sustainable development. The word *ekorozwój* was beginning to be used by Polish environmentalists in the early 1980s. I have not been able to identify the person or organization that invented the word. However, many of those who used the word claimed that the term *ekorozwój* originates from the United Nations Conference on the Human Environment in Stockholm, 1972.

[110] Representatives of the following disciplines were present: agriculture, architecture, economy, ecology, forestry, geography, geology, law, medicine, pedagogy, technology, and urban planning.

[111] In the summary of the conference proceedings, eco-development was defined in the following way: "Thus, eco-development may be understood as the social, technical, management, and cultural progress respecting the requirements of the natural environment. For the realization of eco-development a reconstruction of the way of thinking is necessary; the surrounding world should be seen as a whole and as a system; a balance within nature must be reconstructed which imposes limits and barriers on man" (PKE, 1986: 345). Furthermore: "The aim of eco-development is a better fulfilment of man's physical and psychological needs by the creation of his proper relationship with the natural environment. Eco-development must secure the natural bases of human existence and introduce economic, functional, and aesthetic harmony within the environment in which we live" (*ibid.*: 346).

development where the concept of sustainable development was highlighted.[112]

The PKE sharply rejected the notion that Poland could not allow itself to make environmental protection a priority in times of economic crisis. "Reducing expenditure on environmental protection according to the slogan 'first we will become rich and then we will take care of the environment' did not help. On the contrary, it brought about an economic and moral catastrophe. It also destroyed the Polish nature to such an extent that the existence of the people is endangered" (PKE, 1987: 34). Reality, the PKE argued, must be looked upon as a whole, it cannot be treated in a reductionistic way (*ibid.*: 10). The limited carrying capacity of the environment makes it necessary for people to accept a moderate life style; the consumer society is unacceptable.

The transition to eco-development will imply:

- an economic development within the limits of the carrying capacity of the natural environment;
- an integration of environmental concern into all economic activities;
- use of natural resources in such a manner that they will be available for future generations;
- national self-sufficiency with regard to food, energy, water, and other natural resources; and
- a social and economic development based on small-scale structures.

The realization of these principles will require new criteria for assessing social and economic progress and a fundamental reorientation of the economic sciences.

Economic mechanisms for environment protection should be strengthened in connection with the ongoing economic reform.[113] New investment should be preceded by an environmental impact assessment and the planning process should take social, economic, spatial, and natural factors into account. Environmental problems ought to be addressed at the level where they occur (*ibid.*: 30).

The environmental degradation caused by industry should be addressed by structural changes. Branches "requiring more intellectual innovation", for example, electronics and precision industry, must gradually replace heavy industry. As regards consumption, it was proposed that food packages should contain information about the level of pollution that the food contains. In addition, the PKE pointed to alcoholism, smoking, and drug addiction as serious problems which need to be addressed, and indicated that it was prepared to cooperate with the Ministry of Health to deal with these problems. The main pillars of Poland's future energy policy should, according to the PKE, be energy saving and renewable energy sources.

Nuclear power

At the beginning of 1982, construction began on Poland's first nuclear power plant in Żarnowiec, northwest of Gdańsk. Four reactors of the Soviet VVER 440, type each with a capacity of 465 Megawatts, were to be built. The first reactor in Żarnowiec was

---

[112] Despite the fact that the PKE's analysis in many respects replicates many of the central views of the Brundtland commission, it did not refer to it.

[113] The document refers to reforms initiated by the socialist government.

planned to come into operation in 1991 (World Bank, 1989b).[114]

The plant was part of a large-scale nuclear programme which initially envisaged eleven nuclear power stations. The rationale of the programme was that energy demand was expected to rise and that coal – on technical, geological as well as economic grounds – could not satisfy this demand (*ibid.*). The nuclear plants were to be situated in the northern part of the country. At least ten different places were seriously considered for the location of nuclear power plants (Kozłowski, 1986: 91).

In the first half of the 1980s there was a relatively high social acceptance for the nuclear programme. The debate about the programme was focused on the location of the individual plants.[115]

On the night of 26 April 1986, one of the four reactors at the Chernobyl nuclear power plant in Ukraine exploded and the world's biggest nuclear catastrophe was a fact. Outside the former Soviet Union, Poland was probably more seriously affected by the catastrophe than any other country. In the last days of April 1986 virtually all parts of Poland were covered by fall-out from the accident (Carter and Turnock, 1993: 119).[116] The Soviet cover-up of the tragedy was deemed to fail when the fallout passed the iron curtain.[117]

Below follows an account about the Polish reaction to the accident by Tadeusz Belerski, who was in charge of environmental affairs at the Polish News Agency (*Polska Agencja Prasowa* or PAP) at the time of the accident:

The PAP received the information from Reuter and handed it over to the Central Committee of the PZPR. The press did not receive the news because that was prohibited. During the night between Monday and Tuesday a governmental commission for Chernobyl was created. Tuesday morning I was called up at home by somebody who told me to come to the office of deputy Prime Minister Szałajda. I had no idea what it was all about. I met with the commission and was informed about the accident. This commission had the following members: deputy Prime Minister Szałajda, the Defense Minister, the Minister of Health, the Minister for Internal Trade, a number of representatives from the military, the chief of the civil defense, the main sanitary inspector of the country. There was also the deputy Minister of Health, the director of the Institute for the Mother and Child, and the director of the nuclear

---

[114] VVER is a Russian abbreviation for pressurized water reactors, one of two types of nuclear reactors that were developed in the Soviet bloc. The other type is RBMK, which is a Russian abbreviation for graphite moderated reactors. It was a RBMK reactor that exploded in Chernobyl (Marples, 1986).

[115] For example, the Committee for Spatial Development of the Country (*Komitet Przestrzennego Zagospodarowania Kraju*) of the Polish Academy of Sciences pointed out that the planned power station in Karolewo would be situated within a landscape park. Moreover, the siting would be against the directives for areas of ecological hazard defined by the National Socio-Economic Plan for 1983-85 (*ibid.*: 93). In places like Karolewo and Kujawy, the local population expressed its worries for the future (Mielczarek, 1983).

[116] The consequences of Chernobyl in Poland were still evident in the 1990s. In 1994, Polish authorities issued public health warnings against the consumption of game and mushrooms from several areas of high concentration of caesium 137 (Millard, 1998).

[117] Increased levels of radioactivity outside the Soviet bloc were first discovered at the Forsmark nuclear power station in Sweden in the morning of 28 April. The immediate reaction by the staff at this plant was that an accident had occurred at Forsmark. The personnel were evacuated. In the evening, the same day, the Soviet news agency Tass informed the world that a nuclear accident had occurred in Chernobyl (Karlsson, 1996).

institute in Świerk.[118]

The commission, which met regularly in the weeks after the catastrophe, was a high-level task force created to deal with a problem that had both environmental and political implications for Poland. Hence, the Polish political leadership was not only fully aware but also truly worried about the dimension of the disaster. However, this approach was not reflected in the official response. In the first few days the general public was denied any information. Three days after the accident the PAP sent out a short news message, which noted that an explosion had occurred at Chernobyl and that the situation was under complete control.[119] The government hesitated long about whether to take protective measures or not. This dilemma was finally solved by a half-measure that suggested that children under sixteen in certain regions were given iodine for protection against thyroid cancer.[120] It is noteworthy that the governmental commission on Chernobyl seriously considered evacuating all young people under sixteen living in the contaminated areas in eastern Poland.[121]

Even though Chernobyl shocked the party-state, the catastrophe did not lead to a reconsideration of the Polish nuclear programme. The construction of the nuclear power plant in Żarnowiec continued and Klempicz, northwest of Poznań, was selected to become the place for the second nuclear power plant. In addition, it was intended to use the old German underground fortress at Międzyrzecz, west of Poznań, for storage of nuclear waste (World Bank, 1989b: 13).[122]

Public opinion was greatly influenced by Chernobyl. The opposition to nuclear power made itself known in numerous demonstrations around the country. Most of these demonstrations were held in the areas most affected by the nuclear programme, that is, Poznań and Gdańsk, as well in Kraków, where the headquarters of the PKE was located. The club "made the abandonment of plans for the development of nuclear energy in Poland one of the planks of its programme of alternative development for the country" (*ibid.*). The PKE claimed that international experience showed that energy produced at nuclear power plants was the most expensive one. The government should create a commission given the task of assessing the economic rationale of nuclear power compared to other energy sources. A decision to launch the nuclear programme should not be taken without approval by the people in a *referendum on a fully democratic basis.*

The rationale of the nuclear programme was openly debated at public meetings as well as in newspapers and magazines. Some energy experts, like Włodzimierz Bojarski (1988) of the Polish Academy of Sciences, sought to show that nuclear produced electricity was much more expensive than electricity produced by coal fired

---

[118] Personal communication with Tadeusz Belerski, 1995.

[119] However, thanks to Radio Free Europe most Poles were informed about what had really happened in Chernobyl.

[120] This was done in May 1986. It is generally considered that the iodine was given too late to protect children. The additional dose of iodine may even have been harmful to them.

[121] Personal communication with Tadeusz Belerski, 1995.

[122] These plans led to local mass manifestations. In 1987 one quarter of the 20,000 inhabitants of the town demonstrated against the project under the slogan "we want to live" (Waller and Millard, 1992: 166).

power plants fully equipped with modern flue gas desulphurization (FGD) devices. Hence, the opponents used both economic and environmental arguments against the nuclear programme.

As regards the political leadership, Chernobyl did not change the energy policy but the disaster reinforced their awareness about the shortcomings of the existing socialist system. This is aptly formulated by Witold Maciejewski of the PKE as follows:

Chernobyl was a turning point in their way of thinking about the power apparatus. The catastrophe made many people in the political elite afraid that they had contributed to the creation of a monster they could not control, a political and economic monster, and an energy monster as well.[123]

4.2.3. Interactions

In the 1970s the environmental policy process had been static: virtually all proposals originated from the government, and the policy-makers had constituted a relatively closed circle. The general public did not challenge official policy; there was no open debate and no conflicts. This changed in 1980-81 when the wants and demands from the 1970s were elevated to the status of issues and when a public agenda came into being.

The environmental policy network, including the environmental movement, was much less affected by martial law than Solidarity. Furthermore, in the daily press, environmental news were subject to less censorship than most other news. It appeared that "the environment" had gained a special position in the new political setting, a tendency that became increasingly evident in the 1980s.

During and after the Solidarity period, Polish society began to confront its old problems in a new way. New insights were gained into what the problems were about. New actors became involved in the policy process.

The official policy-makers were not immune to this process, a fact that is illustrated by a statement made by Ludwik Ochocki, a deputy minister at MGTAiOŚ, in *Aura* at the beginning of 1981:

There are many problems which ought to be looked at differently today in comparison to earlier days. Due to increasing public pressure, we must now act more offensively, in accordance with the public will (...). Our environmental problems are not like the Egyptian plagues, something which cannot be dealt with. In the 1970s, environmental problems were treated as phenomena (...). In the 1980s, the state cannot shirk its responsibility for the environmental situation, cannot delay decisive action to address the accumulated problems. We can no longer apply old solutions (Domański, 1981a).

Hence, Ochocki admitted that the activation of the society had brought about new insights into the problems and solutions. In other words, he discovered a connection between public pressure and the quality of environmental policy.

During the course of the 1980s, more and more actors within the environmental policy network became increasingly aware that they were dependent on each other. The Ministry of Environmental Protection and Natural Resources (*Ministerstwo Ochrony Środowiska i Zasobów Naturalnych* or MOŚiZN), formed in 1985, clearly looked upon the environmental movement as its friend rather than its foe. For

---

[123] Personal communication, 1995.

example, Waldemar Michna,[124] Poland's Minister of Environment in 1987-88, claims that it was much easier for him to cooperate with the PKE than with the head of the communist party. Wacław Kulczyński, a deputy Minister of Environment in the second part of the 1980s, remembers that "from 1985 the environmental movement demanded contacts and cooperation. We could not avoid it because we had the same goals".[125] The pressure exerted by the environmental movement actors increased the possibilities for the environmental administration to have influence in general economic and sectoral policies. This was particularly evident in Kraków. According to Bronisław Kamiński, the director of the voivodship environmental protection department in Kraków between 1980 and 1988, pressure exerted by the PKE and other organizations was instrumental for the authorities' possibilities to close the aluminium smelter in Skawina in 1981. He claims that without the environmental movement his position within the voivodship authority would have been much weaker.[126]

During the latter part of the 1980s the Ministry of Environmental Protection and Natural Resources gave clear signals that it accepted a greater role for public opinion in defining the context of environmental policy. As Józef Kozioł puts it: "the doors of the Ministry were opened".[127] For example, in 1985, the Ministry invited the PKE to take part in the elaboration of the long-term environmental protection programme to the year 2010. According to Waldemar Michna, the PKE did not put forward any proposals that fundamentally differed from the views of the Ministry.[128] Furthermore, several persons from the PKE were invited to become members of the State Committee for Environmental Protection. When Koziół visited the Swedish government in March 1989 he was accompanied by the chairman of the PKE, Stanisław Juchnowicz.[129]

By the mid-1980s, six environmental organizations – the League for Nature Protection, the League to Combat Noise, the Polish Tourist Association, the Polish Angling Association, the Polish Hunting Association, and the Polish Ecological Club – were authorized to participate in civil proceedings to sue polluters (Cole, 1998: 77). Hence, they were invited by the environmental authorities to participate in environmental law enforcement.[130]

The establishment of the Ministry of Environmental protection and Natural Resources in 1985 was partly a result of pressure exerted by the environmental movement. In 1981 Stefan Kozłowski of the PKE had complained that the

---

[124] Personal communication, 1995.

[125] Personal communication, 1995.

[126] Personal communication, 1995.

[127] Personal communication, 1995.

[128] Personal communication, 1995.

[129] Kozioł recalls that he "urged Juchnowicz to tell the Swedish government what he wanted" (personal communication, 1995).

[130] For a number of economic and political reasons environmental associations were only occasionally participating in civil proceedings against polluters in the 1980s.

environmental management system was suffering from an organizational diffussion. He called for a ministry entirely devoted to the task of environmental protection. The same idea was put forward by the League for Nature Protection (Kozłowski, 1981). The fact that there were environmentalists who advocated the creation of a ministry in charge of environmental protection suggests that they were prepared to work for environmental improvements within the existing political framework.

What has been said above suggests that Poland's environmental policy process in the 1980s took the appearances of the interactions that are typical for societal corporatism.

Throughout the 1980s there were somewhat more interactions between the environmental policy-makers and the decision-makers in charge of economic development than there had been in the 1970s. As was shown earlier in this chapter, the consequences of Chernobyl were dealt with at the highest political level. A few closures of polluting industries occurred and some projects (mining projects, nuclear power) were postponed.

However, the general picture was not very much different from the situation in the 1970s. The industrial community had no reason to involve itself in the debate or in the policy adoption process because of the conviction that any environmental policy that was adopted could be easily blocked at the implementation stage. Without financial resources, the industry argued, little could be done. Most of the interaction between the environmental and industrial subsystems in the 1980s tended to be a matter of bargaining about subsidies for environmental protection investments. This explains why industry did not bother to form its own advocacy coalition within the environmental protection subsystem.

4.3. Policy adoption

Shortly after the parliament had enacted the Statute on the Protection and Shaping of the Environment in 1980, the Council of Ministers issued ordinances about air protection, noise protection, waste management, cooperation between territorial authorities and other environmental authorities, principles for fines, changes in the fee system, principles for the establishment of the Fund for Environmental Protection and the State Inspectorate for Environmental Protection, and definitions of activities for the State Council for Environmental Protection (Aura, 1980d). The long and arduous process, which had begun in the mid-1970s, to provide the legal basis for the establishment of a coherent environmental protection system, was finalized.

This section will show that a main feature of environmental policy in Poland in the 1980s was the search for mechanisms to enhance the integration of ecological concerns into the general economic policy of the country. It was more and more realized that the approach of the 1970s, that is, application of end-of-pipe technologies, was inadequate and that preventive measures and structural changes in the economy were needed.

Areas of ecological hazard

A decree of the Council of Ministers at the beginning of 1983 officially designated twenty-seven "areas of ecological hazard" comprising approximately 11 per cent of the country's territory and 12.9 million people, i.e. 35 per cent of the population

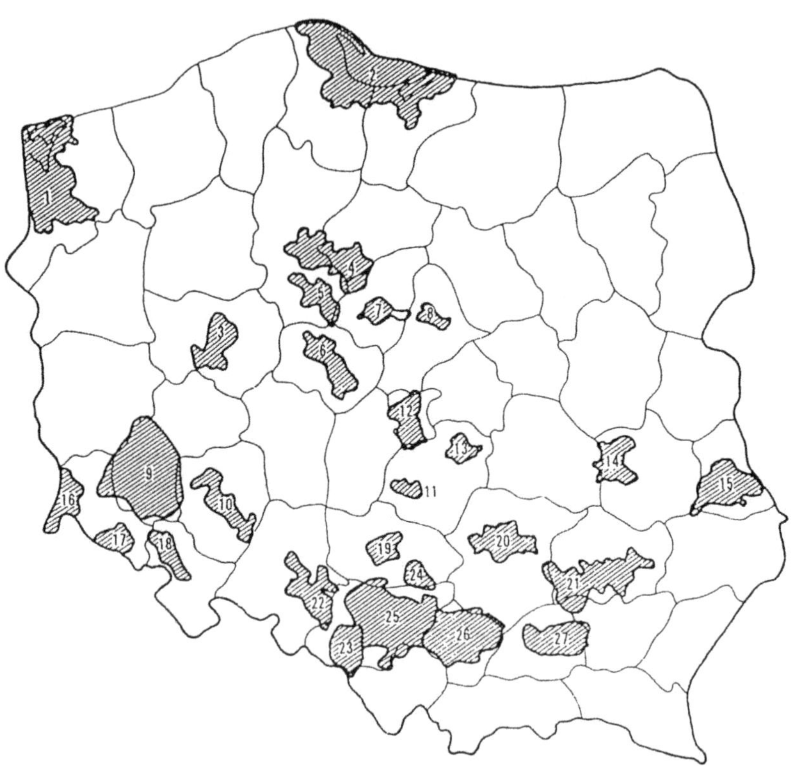

1. Szczecin
2. Gdańsk
3. Poznań
4. Bydgoszcz-Toruń
5. Inowrocław
6. Konin
7. Włocławek
8. Płock
9. Legnica-Głogów
10. Wrocław
11. Bełchatów
12. Łódź
13. Tomaszów
14. Puławy
15. Chełm
16. Turoszów
17. Jelenia Góra
18. Wałbrzych
19. Częstochowa
20. Białe Zagłębie
21. Tarnobrzeg
22. Opole
23. Rybnik-Myszków
24. Zawiercie
25. Górny Śląsk
26. Kraków
27. Tarnów

Figure 4.1. The twenty-seven areas of ecological hazard. Source: GUS, 1991.

(Figure 4.1).[131] The twenty-seven areas were divided into three categories (World Bank, 1989c: 29):

1. "Areas of ecological disasters": the Gdańsk Bay, Kraków, Legnica-Głogów, Rybnik, and the Upper Silesian industry area (*Górnośląski Okręg Przemysłowy*).

[131] The preparatory work of ecological maps had started in 1980 at the Department for Long-Term Spatial Planning at the State Planning Commission. The work was led by Andrzej Kassenberg and Czesława Rólewicz and involved about fifty people. The work was completed and presented to the environmental commission of the parliament in 1982 (personal communication with Andrzej Kassenberg, 1995).

2. Eighteen regions "experiencing serious environmental damage from multiple sources".
3. Four regions "experiencing serious environmental damage primarily from air pollution".[132]

The factors that were taken into account in the elaboration were air quality, water quality, condition of the soil resources, the state of the forests, and general factors of nature and landscape protection. It was shown that certain parts of Poland experienced environmental degradation that posed a threat to both nature and human health. The designation of the twenty-seven areas of ecological hazard increased substantially the awareness of the need to integrate environmental concerns into national economic policy.

Defining and mapping the zones showed for the first time how the state of the environment is related to policies of industrial concentration and the reliance in industry on so-called "dirty technologies". It also made clear that countermeasures cannot be limited to direct environmental protection as such but must be made to include policies for technological changes in industry, changes in industrial structure, and adoption of alternative patterns of development in affected areas. This means that installation of pollution control equipment, recultivation of soils, and reafforestation of degraded forest lands must be accompanied by broadly focused regional economic development planning that incorporates environmental protection (World Bank, 1989c: 41).

The designation of areas of ecological hazard had a direct and immediate impact on strategic economic policy-making. Most importantly, a *Sejm* resolution on the 1983-85 socioeconomic plan called for a ban on building or expansion of harmful industries in the five areas of ecological disaster. The twenty-seven areas were referred to in all major environmental policy documents throughout the 1980s.

The Statute on Land-Use Planning

A typical feature of environmental policy in Poland in the 1980s was the upsurge for land-use planning. The Statute on Land-Use Planning (*Ustawa o Planowaniu Przestrzennym*) of 1984 made Environmental Impact Assessment mandatory for projects "especially harmful for the environment and human health" (World Bank, 1989f).[133] (The EIA decree of 1985 contained more details on which activities should be seen as "especially harmful".[134])

---

[132] In addition to the twenty-seven areas, sixty cities with environmental problems were identified. Together these towns and areas were inhabited by half of the Polish population (personal communication with Andrzej Kassenberg, 1995).

[133] The work to elaborate EIA methodologies had started as early as in the latter part of the 1970s. The Institute for the Shaping of the Environment (*Instytut Kształtowania Środowiska*) in Katowice, in cooperation with the UNDP and the WHO, performed a study entitled "Introduction of Environmental and Health Impact Assessment Procedures into Planning and Decision-Making in Poland". Within the framework of this programme two EIAs were published in 1982 for two pilot areas, the copper mining area in Legnica-Głogów and the southeastern part of Upper Silesia (Kozłowski, 1986: 9).

[134] With respect to air pollution; water pollution; removal of water in certain areas; soils, agricultural plants and forests; transmission of electricity; noise; production and storage of waste; production of electromagnetic field; and production of highly radioactive substances. EIAs were to be

The Statute on Land-Use Planning increased the opportunities for local government to influence physical planning in general and siting of new economic activities in particular (Baszkiewicz, 1985). It also required that "variant principles" for land-use planning be elaborated and that the general public be consulted in the planning process (Aura, 1984).[135]

The Plan for Territorial Development to the Year 1995

The 1986 Plan for Territorial Development of the Country to the Year 1995 (*Założenia Planu Przestrzennego Zagospodarowania Kraju do 1995 Roku*), elaborated by the State Planning Commission, was an attempt to integrate social, economic, environmental, and other concerns into the regional development (Komisja, 1986: 8). The document identified a number of factors that may impede the socioeconomic development of the country of which five were directly linked to environmental issues (*ibid.*: 71-72). The document announced that certain types of economic activities were going to be prohibited in the most polluted and sensitive areas (*ibid.*: 66).

The Plan for the Territorial Development of the Country was a precedent-setting move. As the World Bank (1989f: 33) observed:

> The Plan is innovative in terms of centrally planned systems in that it is a two-variant mechanism. Some observers in Poland consider this to be a significant departure from traditional planning practice, challenging as it does the monolithic style of central planning and presenting a useful social innovation for other socialist countries to apply.

The National Environmental Programme to the Year 2010

At the PZPR's Ninth Extraordinary Congress in 1981 the necessity to elaborate a new comprehensive national programme for environmental protection was acknowledged (World Bank, 1989e: 5). Four years later, in June 1985, the parliament gave the Ministry of Environmental Protection and Natural Resources the task of elaborating a new long-term national environmental programme (Budnikowski *et al.*, 1988: 115). The work resulted in the 1988 National Environmental Programme to the Year 2010 (*Narodowy Program Ochrony Środowiska Przyrodniczego do roku 2010*).[136] It was

carried out to: (1) determine the existing environmental situation at the proposed site; (2) determine the probable impact of the proposed investment; (3) mitigate negative effects on the environment; and (4) determine costs and benefits of the investment (World Bank, 1989f: 26-28). According to Andrzej Kassenberg (personal communication, 1995), the EIAs were "quasi-environmental impact assessments because they were only performed inside the ministry".

[135] It should be stressed that the upsurge of land-use planning in the mid-1980s did not appear from thin air. As was shown in Chapter 3, Poland's first land-use planning law dates back to 1961. Already shortly after the second world war Poland established an agency responsible for spatial planning (*Centralne Biuro Planowania Przestrzennego*). In 1947 this agency produced the first nation-wide spatial plan in Poland (personal communication with Andrzej Kassenberg, 1995).

[136] The preparatory work involved a large number experts from the Polish Academy of Sciences, the State Planning Commission, certain ministries, voivodships, non-governmental organizations (the PKE was invited to all major sessions during the elaboration of the programme) and approximately fifty research and project bureaus. The experts produced twelve different background reports that provided the analytical framework for the programme.

the last environmental programme of socialist Poland.

The tone of the programme was open and frank: there was a clear-cut recognition of the disastrous environmental situation and the state's failure to protect the environment. Thus, the approach to environmental problems was fundamentally different compared to the one of the long-term programme of 1975. As Waclaw Kulczynski puts it: "In the 1970s the tone was, it is bad but (...), while in 1988 we used adequate words for describing the actual situation. There was no camouflage of the environmental degradation".[137]

The programme was based on a number of interesting and, to certain extent, new assumptions. First, the solution of the ecological problems was seen as requiring *maximal involvement* of the whole society (MOŚiZN, 1988a: 8). The whole decision-making process was to be decentralized and opened up for public participation. Pressure exerted by environmental organizations was recognized as being beneficial for the implementation of the programme. Therefore, the document underscored the importance of providing financial support to environmental organizations as well as establishing direct cooperation with them (*ibid.*: 105).[138]

Second, the policy-makers made clear that environmental protection must not be seen as a separate sector but as an "integrated part of every field of the national economy" (*ibid.*: 77). The accumulation of pollution, particularly in the most industrialized regions, "endangers further social and economic development" (*ibid.*: 15). Environmental protection was not seen as a hindrance to economic development. On the contrary, a rapid environmental recovery, the document argued, would promote tourism, forestry, and production of healthy food, which in turn would provide new incomes for the country.

Third, the programme's ambition was to be *comprehensive* and *interdisciplinary*; that is, it should approach environmental protection "not only from a technical and technological point of view, but also from humanistic, economic, organizational, legal, and educational perspectives" (*ibid.*). Measures to protect the environment should be taken *before* environmental harm arises.[139]

Fourth, the programme emphasized the need to decentralize environmental policy and to adopt it to the varying regional conditions in the country. The role of local and regional authorities was to be upgraded.

Environmental protection activities between 1990 and 2010 were to have the following hierarchical order: (1) reconstruction of heavy industry and the energy sector; (2) reduction of the material and energy intensity in the production processes; (3) broad introduction of clean or semi-clean technologies; (4) increased quality of products; and (5) use of end-of-pipe equipment (*ibid.*: 31).

A central feature of the document was the ambitiousness of its targets for environmental improvements; the percentage reductions for a range of emissions, for example, dust and sulphur dioxide, until the year 2010 were considerable. By the same year, all areas of ecological hazard, including the disaster zones, were to be

---

[137] Personal communication, 1995.

[138] Interestingly, the document noted that some of the NGO activities were "not sufficiently focused", and that closer cooperation between environmental authorities and the environmental organizations would increase the effectiveness of the NGO activities. Hence, the authorities and the NGOs were recognized as being mutually dependent.

[139] In other words, prevention should be the guiding principle.

eliminated.

Poland's international environmental obligations were directly linked to the economic interest of the country:

In order to avoid economic losses and to gain certain economic benefits in international trade, it is necessary to immediately follow the international tendencies and in an early stage solve ecological problems which may harm our country politically and economically if they are addressed too late (MOŚiZN, 1988a: 62).

The document highlighted the shortcomings of the policy instruments that had been in use in Poland during the 1980s. As regards economic instruments, it was recognized that existing economic instruments were inefficient due to several factors. Firstly, the level of the fees and fines was too low to serve as incentive for the polluters to change their behaviour. These instruments were not based on an economic assessment of losses caused by environmental degradation and the actual costs of environmental protection. Secondly, due to inflation, these instruments were rapidly losing their value. Thirdly, the polluters were often exempted from paying the fees and fines, and certain economic entities were not even obliged to pay. The document anticipated a substantial improvement of the situation by the mid-1990s. By then, fees and fines were to be raised so much that it would have become more expensive to pollute than to protect the environment. An immediate measure to be implemented was to replace the existing environmental funds with one fund, the National Fund for Environmental Protection and Water Management (*Narodowy Fundusz Ochrony Środowiska i Gospodarki Wodnej* or NFOŚiGW). In the long run, environmental protection was to become self-financing. After 1995 the financial burden should primarily be incumbent upon the polluters, not the state. Financial means from the state budget should only be used occasionally.

The document argued that there was lack of congruence between the environmental laws: many of them contained contradictory goals, environmental legislation was dispersed over too many acts which made them difficult to enforce; regulation about environmental protection activities within enterprises was insufficient; and the civil and penal codes neglected environmental crimes.

The Round-table in 1989

One of the topics discussed during the Round-table discussions between the government and Solidarity at the beginning of 1989, which led to the re-legalization of Solidarity and ultimately ended the socialist power monopoly in Poland, was environmental policy reform. At the environmental table, the opposition was mainly represented by key persons from the PKE and Solidarity,[140] while the government side consisted of representatives of the Ministry of Environmental Protection and Natural Resources and the official labour union the All-Polish Agreement for Trade Unions (*Ogólnopolskie Poruzumienie Zwiazków Zawodowych* or OPZZ). On both sides representatives from the research community, including the Polish Academy of Sciences, were present. After one month the two parties had agreed upon twenty-eight postulates, formulated in the Protocol of the Sub-Assembly for Ecological Affairs of the Round-table. The ecological protocol clearly reflected the fact that Poland was

[140] Out of fourteen participants on Solidarity's side, five were from the Polish Ecological Club, including the co-chairman Stefan Kozłowski (Górka, 1991).

heading towards a new political system. For example, it was stated that there should be no restrictions in the dissemination of information about the environmental situation; censorship should be abolished; all environmental groups should be guaranteed total freedom to act, protest, and demonstrate; and representatives of environmental non-governmental organizations and the public should have the right to participate in policy-making on central, regional, and local levels. Moreover, "persecution of individuals who pursue ecological activities in the public interest should cease immediately" (Instytut na rzecz Ekorozwoju, 1993: 52).[141]

Eco-development should be the corner-stone in the future social and economic development of the country. Such development, it was argued, is necessary for the ecological balance and a precondition for further social and economic development. The implementation of eco-development requires:

- *integration* of ecological goals both into the social and economic goals and into the physical planning of the country;
- *reconstruction* of the national economy, particularly industry;
- elaboration of *regional plans* for eco-development. Priority should be given to the most polluted areas and the territories with the highest nature values; and
- introduction of mandatory *environmental impact assessments* in all plans.

There were several suggestions showing an increasing interest in environment protection as a way of gaining economic benefits. For example, the creation of a bank for environmental protection was proposed. Efforts were envisaged to reduce the foreign debt by increasing the Polish investment for environmental protection according to the model debt-for-nature swap. This idea was to be promoted jointly by the ministries of environment, finance, and foreign affairs as well as the PKE.

The document stated that raising the general ecological awareness "is a precondition for a successful implementation of the ideas of a eco-development" (*ibid.*: 50). Therefore, the Ministry of Environmental Protection and Natural Resources was to initiate and coordinate an educational programme for the whole society.

Water management was to be improved by the creation of Regional Water Management Boards (*Regionalne Zarządy Gospodarki Wodnej*) corresponding to the hydrological division of the country. The Ministry of Environmental Protection and Natural Resources was to be in charge of all natural elements and therefore needed to incorporate forestry into its sphere of responsibility. Existing environmental legislation was to be thoroughly revised and modernized. For this purpose a committee on environmental law reform was to be established.

PIOŚ should be strengthened and given the right to discontinue environmentally harmful activities and stop the initiation of projects likely to be damaging to the environment. A list of the most environmentally harmful industries should be elaborated. Rapid decisions about how to eliminate the negative impact of these plants should be taken. Certain industrial plants constituting an extraordinary danger for the environment and public health should be closed within a year.

The only issue that caused disagreement was the future of the Polish nuclear programme. The government side was in favour of this programme, while the opposition advocated a non-nuclear future.

---

[141] One of the environmental activists at the Solidarity side of the Round-table, Michał Downarowicz, had been beaten by the police in a street in Poznań as late as in November 1988 (personal communication with Michał Downarowicz, 1995).

Remarkably, all suggestions put forward by the opposition but one (the proposal to cancel the nuclear programme) were accepted by the government side. By and large, most of the twenty-eight postulates reflected the ideas that the environmental movement *and* the Ministry of Environment had advocated in the preceding years.

However, the ecological Round-table was not an unqualified success; the discussions were held in isolation from the groups where the main political and economic issues were brought up. In some groups, for instance the one concerning mining, decisions were made which were considered to be against the policy proposed by the ecological Round-table.[142]

4.4. Implementation

In the 1980s it was possible to discern some signs of improvement in Poland's environmental performance. The environmental expenditures as a percentage of GDP grew from 0.28 in 1980 to 0.8 in 1989 (Cole, 1995b). The Environmental Protection Fund and the Water Management Fund financed most of the environmental protection investments. In 1984, for instance, these funds covered 72 per cent of the expenditures (Górka and Poskrobko, 1987: 181). The funds benefited from a high collection rate of the environmental fees. The collection rate was 97.6 per cent in 1985, 87.5 per cent in 1987, and 88.7 per cent in 1988 (GUS, 1989). More waste water treatment plants were built in the 1980s than in the 1970s. For instance, in 1989, 54 per cent of the industrial plants treated their sewage water, compared to 40 per cent in 1975. Dust emission was reduced from 2.3 million tonnes in 1980 to 1.5 million tonnes in 1989 (Alberski, 1996: 58-61).

PIOŚ exercised an increasing level of controls of the polluters. For instance, in 1984, the checks amounted to 18,000 in 11,000 different industrial plants (Zaręba, 1985). Since 1980 the voivodship authorities had the instrument of closure in their hands. This tool was used on two occasions during the 1980s: the aluminium plant in Skawina outside Kraków in 1981 and Celwiskoza, a big viscose fibre works in Jelenia Góra in 1987 (Jendrośka and Radecki, 1991). Another success was that the planned exploitation of iron ore in a landscape park in Suwałki in northeastern Poland was stopped because of opposition from the local authorities and the Polish Academy of Sciences (Chmielewski, 1988: 56).

However, the overall picture was bleak and the above-mentioned improvements did not match the negative trends. Gaseous air pollution emission increased substantially from 1985 to 1989 (*ibid.*: 62). Lack of advanced technology to solve this problem appeared as a major barrier. Saline waste water discharges from the hard coal sector continued unabated. Very little was done to address the problem of solid waste from the industrial sector. Moreover, the advanced tools for land-use planning were never fully used in the 1980s. According to Andrzej Kassenberg,[143] who worked for the Department for Perspective Spatial Planning at the State Planning Commission in the

---

[142] Andrzej Kassenberg, one of Solidarity's representatives at the ecological Round-table, remembers a brief conversation with Tadeusz Mazowiecki (the Prime Minister in the first Solidarity led government 1989-90) during the Round-table. Kassenberg: "Something anti-ecological is going on in the mining group." Mazowiecki: "I know. But right now something else is much more important". Kassenberg: "Maybe you are right" (personal communication with Andrzej Kassenberg, 1995).

[143] Personal communication, 1995.

1980s, this was related to two factors: the socialist ideology still came first in the decision-making process and enforcement was weak. Moreover, the EIAs did not have a real impact on the development process. As Jendrośka (1992) concludes:

Neither developers nor authorities knew how to perform the EIA, there were no regulations providing requirements as to the content of EIAs, and there were no procedures enabling to evaluate their quality. As a result, the majority of the projects were not made subject to EIA, and even if they were – EIAs were of a very poor quality.

The 1980 Statute conspicuously failed to reach many of its objectives.[144] Sanctions for non-compliance were not enforced. In 1984 PIOŚ revealed that of 7,000 enterprises, 70 per cent had no permits for water pollution and 80 per cent lacked a permit which defined the legal emission of air pollution (Aura, 1985). At the end of the decade the fees rapidly lost their value because of looming inflation. An additional problem during the period was that existing financial resources for environmental protection were not fully used because of lack of equipment.

It should be emphasized that the most important implementation barrier was not technical or economic but political: the environmental policy-makers had virtually no real influence in the politbureau and the central committee of the PZPR which were the fora where the most strategic decisions for the country were taken. In this sense very little had changed compared to the 1970s.

4.5. The Environmental Protection Advocacy Coalition and its belief system

The environmental policy network of the 1980s can be segmented into four main groups of actors: the environmental protection administration, the research community, the environmental organizations, and media.

1. *The environmental administration.* After the enactment of the Statute on the Protection and Shaping of the Environment in 1980 the administrative structure of environmental protection developed considerably. In 1983 environmental protection was removed from MAGTiOŚ, and a governmental Bureau of Environmental Protection and Water Resources Management (*Urząd Ochrony Środowiska i Gospodarki Wodnej* or UOŚiGW) was established (World Bank, 1989d: 28). It was the first central agency in Poland devoted solely to environmental protection. The Bureau was led by Professor Stefan Jarzębski, a metallurgist from Upper Silesia and an air protection specialist, who was given the status of a Minister. In 1985 the Bureau was transformed into the Ministry of Environmental Protection and Natural Resources, still led by Jarzębski. The Ministry was a considerably larger organization than its predecessor and it was authorized to deal with a broader set of issues than UOŚiGW. An important change of the 1985 reorganization was that nature protection was moved from the Ministry of Forestry to the newly created Ministry of Environmental Protection and Natural Resources.[145] (The structure of the Ministry in 1988 is depicted in Appendix 3.) The importance of PIOŚ gradually increased in the

---

[144] As Maciej Nowicki, Minister of Environment in 1991, puts it: "The 1980 Statute was a really good law. But the day-to-day practice was different" (personal communication, 1995).

[145] Personal communication with Józef Kozioł, 1995.

1980s. By the end of the decade it relied on a network of ten regional departments with some 300 employees.[146] Other important actors in the environmental administration were the voivodship departments for environmental protection and water management,[147] OBiKŚ,[148] SANEPID, the Environmental Protection Fund, and the Water Management Fund. There was also a parliamentary commission for environmental protection and natural resources and two advisory bodies to the Ministry of Environment, *viz.* the State Committee for Environmental Protection and the State Committee for Nature Protection. In 1989 the Ministry of Environment merged the Environmental Protection Fund and the Water Management Fund into one fund: the National Fund for Environmental Protection and Water Management. It can also be noted that in the early 1980s, the Main Statistical Office created a special department for environmental issues.

2. *The research community.* The actors affiliated with the scientific sector grew in importance in the 1980s and became one of the most vocal environmental advocates. Existing scientific organizations increased in importance at the same time as new bodies were created. The research community produced ample evidence about the environmental crisis, which was presented in a large number of reports, and frequently organized seminars and conferences. The scientists also had considerable leeway in conducting critical analysis of the deficiencies of environmental policy. The most important actors were: the Committee on Environmental Engineering, the Committee of Man and the Environment, the Committee for the Spatial Development of the Country, and the Institute for Basic Technical Problems (all under the auspices of the Polish Academy of Sciences), the environmental protection committee of the Main Technical Organization, departments for environmental engineering at various technical universities (such as in Warszawa), the Scientific Association for Environmental Law (created in 1987) and scientific research institutes for environmental protection (formally belonging to the Ministry of Environmental Protection and Resources) in Katowice, Gdańsk, Warszawa, and Wrocław.

3. *The environmental organizations.* LOP remained by far the largest environmental organization in the 1980s. LOP was radicalized during the Solidarity period but still did not play a major role in the policy debate. The central NGO in this debate was the PKE. The NGO-community was further strengthened after the Chernobyl disaster in 1986. The economic and political liberalization in 1988 was followed by the emergence of new groups such as the Green Federation and the Polish Ecological Party.

4. *Media.* A fourth important group of actors was the media, especially various magazines and the daily press. Popular magazines such as *Aura* and *Przegląd Techniczny* reached a larger and larger audience. The articles published in these, and many other magazines and journals, had a frank and open tone; scientific evidence was made available to the public and a debate was allowed in which criticism against the government could be articulated. The leading newspaper with respect to environmental issues was *Życie Warszawy* that regularly published articles by Iwona Jacyna, an influential environmental journalist. The newspaper frequently published

---

[146] Personal communication with Zbigniew Kamieński, 1995.

[147] In 1988 these departments had 1,236 employees (GUS, 1989).

[148] In 1988 OBiKŚ had 2,541 employees (GUS, 1989).

lists from readers who wanted to focus attention to a particular problem. Environmental issues were also a common topic in various political, economic, and cultural magazines such as *Polityka*, *Kultura*, *Argumenty*, *Tygodnik Demokratyczny*, *Prawo i Życie*, *Literatura*, and *Życie Gospodarcze*. Also radio and television devoted considerable attention to the environment via documentaries, discussions, and lectures (Ochocki, 1985). At the beginning of the 1980s journalists specializing in environmental issues created their own organization, *Ekoś*.

From what has been said above, it follows that the nascent environmental subsystem of the 1970s evolved into a relatively mature one in the mid-1980s. At this time Poland had established specialized subunits within all levels of government to deal with the topic and also independent interests groups for environmental protection were in place. The key events in this process were the establishment of the Ministry of Environmental Protection and Natural Resources and the PKE.

The maturation of the environmental policy subsystem coincided with the emergence of what could be labelled an Environmental Protection Advocacy Coalition (EPAC) consisting of virtually all of the actors outlined above. It fulfilled the fundamental criteria of an advocacy coalition both with respect to the degree of coordinated action (see Section 4.2.3) and shared beliefs.[149]

The environmental policy agreement at the 1989 Round-table gave strong evidence that the belief system of the main actors within the environmental policy subsystem was the same at the eve of systemic change.

Deep core beliefs

The actors within the EPAC reflected more seriously on the roots of the environmental problems than they had done in the 1970s. This implied, *inter alia*, that socialism was no longer seen as a panacea for environmental problems.

Policy core beliefs

The proper relationship between environment and economy. Environmental protection was no longer treated as a policy area within its own right. The necessity of structural changes in the economy and integration of environmental concerns into the economic decision-making process was emphasized in a number of documents produced by the environmental administration, the scientific sector, and the environmental movement. The notion of eco-development was advocated, first by the environmental movement and later on by the rest of the coalition. The actors in the EPAC agreed that environmental degradation constituted a barrier for further economic development of the country. There was a common conviction that industrial production was endangered by the excessive pollution loads. The coalition frequently referred to calculations that showed that the GDP losses of environmental degradation were many times higher than the expenditures for environmental investments. The EPAC also became increasingly convinced that environmental protection is a profitable endeavour, both for the country and for the polluters. In this respect the EPAC's belief system had similarities with the ideas of ecological modernization that emerged in

[149] This conclusion is supported by a majority of the interviewees.

western Europe in the 1980s.[150]

Whose welfare counts? The EPAC's concern was the welfare of all citizens. In the 1970s the welfare of the working class had the highest priority.

The seriousness of environmental problems. Within the EPAC there was considerable agreement that the ecological situation had reached a catastrophic dimension. Poland was recognized as one of the most polluted countries in the world.[151] Virtually all official and unofficial policy documents acknowledged the seriousness of the threat to the country's environment and to the health of its citizens.[152] In certain environmental documents the problems were framed as a threat to the survival of the Polish nation, culture, and people (see, for example, Zimny, 1988).

Basic cause of environmental problems. It was argued that environmental degradation in Poland first and foremost was a result of inadequate economic policies (Budnikowski *et al.*, 1988: 110) and that the system had emphasized production goals at the cost of all other societal goals, including environmental protection. Insufficient attention was paid to the prevention of pollution.

Proper scope of governmental vs. market activity. In the absence of market economy this issue was not addressed in the official discourse.

Proper distribution of authority. The EPAC was a strong advocate of a far-reaching decentralization of the environmental management system.

The choice of policy instruments. The policy actors believed in an increased role for economic incentives (most notably fees and fines) and for tools such as environmental impact assessment and land-use planning to integrate environmental concerns into the decision-making process.

Method of financing. There was support for the idea that the main financial burden should be incumbent on the polluters.

Ability of society to solve the problems. It was implicitly assumed that it was possible to improve the environmental situation within existing system.

Who should participate? It was widely believed that the possibilities for public participation in the decision-making processes should increase. Also the role of environmental NGOs should be expanded.

Secondary aspects beliefs

Institutional aspects. In the 1970s political loyalty was seen as more important than any other personal qualities. The EPAC required that the environmental administration be characterized by professionalism.

Priority problems. In the 1970s water management and nature protection were the priority issues for the environmental policy-makers. In the 1980s these issues remained important for the EPAC but the coalition also highlighted issues such as

[150] The ideology of ecological modernization has reconceptualized the relationship between economy and environment. A central element of this ideology is that environmental protection is not seen as a burden upon the economy but a potential source for future growth (Weale, 1992: 76).

[151] In an assessment of the air quality situation in Katowice voivodship Jan Kapała, an air protection expert of the Polish Academy of Sciences, claimed that the health threat to the population in this region was exceptional in a European perspective. In his view, the city Zabrze had "probably the worst air quality in the world" (Kapała, 1983: 73).

[152] See for example MOŚiZN (1988a), Wójcik (1985), and PKE (1987).

public health and air pollution.

The EPAC was a large and rather amorphous coalition with a not fully coherent belief system. The environmental movement was more radical in many respects than the officials within the environmental agencies. For example, the NGO-community, together with many scientists, was against nuclear power, while large segments of the environmental administration tended to defend this energy source. Moreover, the PKE was fully committed to the idea of eco-development while the official policy-makers only occasionally framed the issue of environment and development in this way. Furthermore, within the coalition, there were diverging beliefs about the *raison d'etre* of the socialist system; there were both friends and foes of the existing political order. However, the approach taken here is that presence of diverging opinions is in line with hypothesis 10 of the ACF which states that "within a coalition, administrative agencies will usually advocate more moderate positions than their interest group allies" (Sabatier and Jenkins-Smith, 1999: 17).

I shall not deny the fact that my conclusion that the PKE and the Ministry of Environment formed the core of an Environmental Protection Advocacy Coalition in the mid-1980s may seem strange and even controversial to some analysts. Żylicz, for example, argues that "the Polish Ecological Club and the academic community were acting as a coalition more or less against the Ministry of Environment which was part of the establishment of the rest of the government. The Ministry of Environment was an outreach of the Ministry of Industry".[153] In this chapter I have shown that there was agreement between the PKE and the Ministry on at least four policy core items: the overall seriousness of the problems, the choice of policy instruments, approaches to public participation, and the proper relationship between environment and economy. However, Żylicz is certainly right in one respect: the first Minister of Environment, Stefan Jarzębski (1985-87), was strongly opposed to the whole concept of designating areas of ecological hazard and was generally seen as an "anti-environmentalist" by the environmental movement.[154] Therefore, he did certainly not belong to the EPAC. No doubt, the EPAC consolidated after he had left the Ministry.

Policy-oriented learning

The most important learning events that occurred *within* the EPAC in the 1980s were the following. First, eco-development appeared as the major goal for environmental policy. This goal was first endorsed by the PKE in the mid-1980s and by the end of the decade there was widespread support for eco-development within the EPAC. Second, the EPAC discovered that environmental protection may be a profitable endeavour. Third, the EPAC discovered various tools for integrating environmental concerns into the economic decision-making process. Among these tools were land-use planning and Environmental Impact Assessment.

In the 1980s there were a number of factors that were conducive to policy-oriented learning within the environmental subsystem. First of all, the special position of the environmental issue gave the environmental policy actors more political freedom than other policy actors. Throughout the 1980s, environmental issues were covered well by the media, particularly the daily press, and environmental news was less censored

[153] Personal communication, 1995.

[154] Personal communication with Andrzej Kassenberg, 1995.

than other news. Furthermore, the main actors within the environmental policy subsystem were allowed to interact relatively freely. An additional factor that facilitated policy-oriented learning was the presence of professionalized fora. In the People's Poland, as elsewhere in the Soviet bloc, the academy of sciences had a strong and relatively autonomous position. Within the Polish Academy of Sciences, the environmental committees served as professionalized fora that admitted participation on the basis of professional training. Such fora are hypothesized by Sabatier and Jenkins-Smith (1993) to be conducive to policy-oriented learning. Also various environmental journals and magazines, such as *Aura* and *Przegląd Techniczny*, may be seen as important fora for policy-oriented learning.[155] A further driving force for policy-oriented learning in the 1980s was the increased interaction with western governments and organizations. The environmental protection subsystem was less restricted than other subsystems in developing international cooperation. An agreement on cooperation in the field of environmental protection between the Polish Ministry of Environment and the US Environmental Protection Agency (EPA) was signed in 1987. The agreement involved among other things cooperative research by teams of Polish and American scientists. Similar agreements were signed with the Netherlands and Sweden in 1987 (World Bank, 1989f: 5). The same year the Institute for European Environmental Policy in Bonn and the Institute of Environmental Engineering in Zabrze entered a joint research programme (World Bank, 1989e). Poland actively participated in the work of the East European Programme of the International Union for Conservation of Nature (IUCN) that began its operations in 1988. In the late 1980s Poland was a party to thirty international conventions and cooperated with the UNEP, the UNECE, the FAO, the WHO, and the UNESCO (World Bank, 1989f: 6). Hence, learning by imitation in the 1970s became learning by interaction in the 1980s. Lastly, learning may have been facilitated by the fact that many actors in the environmental protection subsystem simultaneously represented more than one "estate". This facilitated a quick and effective communication between, for example, the environmental movement and the scientific sector.

4.6. Summary and conclusions about policy change

In the 1980s Poland relied primarily on market mechanisms for environmental protection, which was an important change compared to the 1970s. Several policy documents were approved that emphasized the need to integrate environmental concerns into the economic and developmental policies. At the ecological Round-table in 1989 the government as well as the opposition endorsed the concept of eco-development (the Polish word most commonly used for sustainable development).

Environmental performance improved slightly in the 1980s. But the general picture from the 1970s had not changed: industrial pollution was still out of control. Economic instruments were ineffective in the non-market environment and the embryonic enforcement mechanisms were not effective in a setting where the rule of law still did not apply.

In the mid-1980s Poland established a mature environmental policy subsystem. The Ministry of Environmental Protection and Natural Resources (formed in 1985), the

[155] Together with many other environmental journals specializing in, for example, air protection, water management, and nature conservation.

independent Polish Ecological Club (created in 1980) together with environmentally oriented scientists within the Polish Academy of Sciences, constituted the core of an Environmental Protection Advocacy Coalition (EPAC) which appeared in the middle of the decade. The EPAC highlighted openly the shortcomings of environmental policy and called for improved integration of environmental concerns into the economic policy and regarded the involvement of ministries other than the environmental as essential. The discourse indicated that socialism no longer was seen as a panacea for addressing environmental problems.

Policy change was in the 1980s *not* a result of competition between various coalitions but rather the result of new ways of interaction and policy-oriented learning, in a rather corporatist manner, between the actors within the EPAC. These processes were sparked off by two major external events: the creation of Solidarity in 1980 and the nuclear catastrophe in Chernobyl in 1986.

5. NATIONAL ENVIRONMENTAL POLICY IN THE 1990s

Chapter 5 starts with an introduction containing the main external events of the 1990s. Section 5.2 deals with the policy process. It focuses on actors, issues, and interactions between the actors. Section 5.3 is about policy adoption, especially the national environmental policy programme of 1991 and the approximation of Polish environmental law to EU directives. Section 5.4 focuses on implementation that includes financing, enforcement, and environment and development. Section 5.5 identifies advocacy coalitions and their belief systems and discusses the role of policy-oriented learning. Section 5.6 draws conclusions about the main driving forces for policy change and summarizes the chapter briefly.

5.1. Introduction

The semi-free elections of June 1989 resulted in an overwhelming victory for Solidarity. After PZPR repeatedly had failed to create a new governmental coalition in the summer of 1989 the task was eventually given to Solidarity. On 24 August, Tadeusz Mazowiecki, one of the most prominent figures within Solidarity, became the first non-communist Prime Minister of post-war Poland. In September he managed to form a Solidarity-led government in coalition with PZPR, the Democratic Party (*Stronnictwo Demokratyczne* or SD), and the Peasants' Party (*Zjednoczone Stronnictwo Ludowe* or ZSL).[156] In December 1989 the parliament ended the existence of the socialist state, the People's Republic of Poland (*Polska Rzeczpospolita Ludowa* or PRL). A democratic state, based on the rule of law, was declared (Cole, 1998: 185) and the name of the country was changed to the Republic of Poland (*Rzeczpospolita Polska* or RP) (Węcławowicz, 1996: 27).

Poland's political and economic transformation was speeded up by the geopolitical changes in eastern Europe in the early 1990s: the end of the Warsaw Pact Treaty (July 1991), of the Council for Mutual Economic Aid (CMEA) (December 1991), and of the Soviet Union (December 1991) (Węcławowicz, 1996: 177).

At the time of systemic change the Polish economy was in a miserable state. Poland was plagued by hyperinflation,[157] intense shortages, a growing black market, and falling industrial production (Sachs, 1994: 28). In addition to this, the economy suffered from a number of structural problems: Poland was over-industrialized; the country had a large and inefficient agriculture sector; most of the economy was owned by the state; there were few small and medium-sized enterprises in the industrial sector; and Poland's trade was excessively directed toward to the CMEA area (*ibid.*: 12-13).

---

[156] SD and ZSL were two small parties that had been allowed to exist in People's Poland. After systemic change the ZSL changed its name to PSL (*Polskie Stronnictwo Ludowe*).

[157] Inflation was especially high in the second part of 1989. In October that year the monthly inflation rate peaked at 54 per cent (Sachs, 1994: 40).

The first few years of the transition were a period of economic retrenchment. Poland lost 18 per cent of GDP in 1990-91 (Balcerowicz, 1995: 301) and industrial production went down by one fourth in 1990 compared with 1989 and another 10 per cent in 1991 (Sachs, 1994: 74). In addition, unemployment appeared as a new phenomenon: from the beginning of 1990 until the end of 1992 unemployment increased from zero to more than 13 per cent (Balcerowicz, 1995: 301). In mid-1998 the unemployment rate had been reduced to 10.2 per cent (*Rzeczpospolita*, 1998c).

On 1 January 1990 the Polish government launched a programme to stabilize the economy and depart from central planning. The main pillars of this programme, colloquially called the Balcerowicz Plan after the finance minister Leszek Balcerowicz who was in charge of the programme, were the following: a sharp cut in subsidies for production and consumption; an end to essentially all price control; an elimination of central planning; devaluation of the exchange for the zloty; liberalization of foreign trade; and various measures to encourage competition, de-monopolization, and the entry of new firms (Balcerowicz, 1995: 57).

The year 1994 was an important milestone in the Polish transition to market economy in that at the end of that year more than 50 per cent of GDP was in the private sector (Kołodko and Nuti, 1997). In 1993 the Polish economy embarked on economic growth that has lasted since then. Table 5.1 provides an overview over the major trends in the Polish economy in the 1990s.

Table 5.1 Economic statistics for Poland, 1989-96

	1989	1990	1991	1992	1993	1994	1995	1996
Unemployment (percentage of the labour force)	0.0	6.5	12.2	14.8	16.4	16.0	14.9	13.6
Private sector share of GDP (%)	28.6	30.9	42.1	45.4	47.5	53.0	62.0	70.0
The share of agriculture in GDP (%)	11.8	10.3	9.0	6.7	6.6	6.2	7.9	7.7
The share of industry in GDP (%)	44.1	44.9	40.3	34.0	32.9	32.2	32.0	31.5
GNP per capita (in US$) at Purchase Power Parity exchange rate	Na	Na	Na	Na	Na	5,380	6,200	7,000

Source: Kołodko and Nuti, 1997.

The parliamentary elections of 1991 brought twenty-nine parties to the parliament (Sachs, 1994: 113). Between 1990 and 1993 Poland had four different governments[158] and, likewise, four different Ministers of Environment: Bronisław Kamiński in 1989-90, Maciej Nowicki in 1991, Stefan Kozłowski in the first part of 1992, and Zygmunt Hortmanowicz in 1992-93. The parliamentary elections in September 1993 resulted in a governmental coalition between the Peasant's Party and the Left Democratic Alliance (*Sojusz Lewicy Demokratycznej* or SLD), that governed Poland until 1997. The Minister of Environment in this period was Stanisław Żelichowski of the PSL. After the 1997 parliamentary elections, Solidarity's Electoral Action (*Akcja Wyborcza*

[158] These governments were led by Tadeusz Mazowiecki (1989-90), Jan Krzysztof Bielecki (1991), Jan Olszewski (first part of 1992), and Hanna Suchocka (1992-93).

Solidarność or AWS) formed a government coalition together with the Freedom Union.

At the 1990 European Council in Dublin it was decided that a new type of association agreement between the then European Communities and individual countries in central and eastern Europe be created. These so-called Europe agreements aim at associating the newly established democracies with the European integration process on the basis of the four freedoms of movement of the EU and the internal market. The agreements include an asymmetrical opening-up of EU markets, technical and economic cooperation, and approximation of law. The ultimate goal of the association agreements is full integration with the EU. Poland signed its association agreement with the EU in December 1991 that came into force in February 1994.[159] Joining the EU[160] means adopting some 7,000 pieces of legislation comprising policies on, for example, industry, customs duties, commerce, company law, residence conditions, taxation, social issues, regional matters, energy, environmental protection, agriculture, and scientific research (Sachs, 1994). The cost of approximation of environmental legislation to EU directives could constitute as much as 30-40 per cent of total approximation costs (Miljödepartementet, 1997: 36).

An important milestone of the Polish transition was that the country was granted membership in the OECD in November 1996. Poland hereby gained international recognition as an advanced market economy (Kołodko and Nuti, 1997: 7). In May 1997 a new constitution was approved in a national referendum (Jendrośka, 1997: 106).

5.2. The policy process

5.2.1 Actors

Systemic change abruptly changed the agenda-setting patterns that had emerged during the 1980s. The most interesting phenomenon is perhaps the profound transformation of the environmental movement in the 1990s. Firstly, those who joined the green movement in the 1980s just in order to take up the battle against the socialist system have left. At least two environmental organizations, with clear-cut political missions, have been completely dissolved in the 1990s: Wielkopolska Ecological Seminar and the Silesian Ecological Movement. A second trend is that the NGO community has become more and more professional during the course of the 1990s; its knowledge base has increased substantially and many organizations and groups now have well-equipped offices and their own employees. Efforts have also been made to increase the integration of the environmental movement. Since 1993 an annual national coordination meeting for the environmental groups is being held in Kolumna outside Łódź (*Biuletyn Informacyjny*, 1995). However, the Polish environmentalists are rather isolated in Polish society. They have relatively few members and tend to rely on support from western Europe and the USA rather than

---

[159] In 1998 ten countries in eastern Europe had signed association agreements with the European Union.

[160] Poland is not likely to join the EU before the year 2002 (Business Central Europe, 1997).

from Polish citizens.[161]

In the 1989 election campaign, environmental protection was placed third in Solidarity's list of priorities, behind issues such as food and housing (Hughes, 1990: 1). After 1989 the environment gradually lost much of the prominent position on the public and governmental agendas that it had enjoined in the latter part of the 1980s. The examples are legion: the political parties attach little priority to the environment and few of them have elaborated their own environmental policy programmes (with one important exception, the Freedom Union)[162]; the issue was highlighted neither in the campaigns before the parliamentary elections in 1991, 1993, and 1997 nor in the campaigns before the presidential elections in 1990 and 1995. None of the green parties has been close to coming into the parliament (Alberski, 1996). In the parliamentary election in 1991 sixty-three green lists were registered in more than one constituency. Of these three regarded themselves as ecological parties: Healthy Poland (*Zdrowa Polska*), a coalition consisting of the Polish Ecological Party (*Polska Partia Ekologiczna*) and the Polish Greens' Party (*Polska Partia Zielonych*), and the Polish Ecological Party – Greens (*Polska Partia Ekologiczna – Zieloni*). They did not win any seats (Waller and Millard, 1992: 168). To this should be added that the Ministry of Environmental Protection, Natural Resources and Forestry has enjoyed a low status in all governments that have been formed since 1989. In 1994-95 the government seriously considered dissolving the Ministry, together with PIOŚ, in connection with the reorganization of the central state administration (BSE, 1995).[163]

Stefan Kozłowski (1994) speaks about a "political disaster" for the environmental movement.

> Firstly, the environmental movement has become discredited, even ridiculed. Secondly, there is no strong ecological lobby for sustainable development in the parliament. The gains of the 1980s are, to a large extent, wasted. The question thus arises when a credible environmental formation, capable of having a substantial influence on the development of the country, will be created in Poland?

A major reason why environmental issues have lost saliency in Poland in the 1990s may be that the transition period has imposed economic and social burdens on the population.

---

[161] It is worth mentioning that the largest environmental organization in Poland, the League for Nature Protection, has lost about one million members in the 1990s. It had 1,340 million members in 1990 and "only" 358,000 members in 1996. Of these, more than 90 per cent were school children (GUS, 1997). See Gliński (1995) for an extensive overview of the activities of the environmental organizations in Poland in the 1990s.

[162] The Freedom Union has created a nationwide Ecological Forum to promote environmental protection. Since 1991 the party issues an environmental magazine entitled Green Alternative (*Zielona Alternatywa*, 1991).

[163] This idea met with furious opposition from the environmental community and was almost immediately abolished (Alberski, 1996).

5.2.2. Issues

Transport

One of the new issues which has appeared on the Polish environmental agenda in the 1990s is transport. The number of passenger cars per 1,000 people increased from 61 in 1980 to 167 in 1992. With the current trend private cars per 1,000 persons will increase to almost 400 in the year 2010, a level comparable to western Europe (Anderson, 1994). The market share of public transportation in Poland decreased about 34 per cent between 1986 and 1993 (Ministry of Environmental Protection, 1998). These trends imply that private transport is contributing with more and more air pollution.[164]

The programme for the construction of a national motorway network (expected to comprise about 2,000 km) within the next twenty years has met with fierce opposition from a number of environmental organizations, most notably the Polish Ecological Club, the Green Federation, and the Institute for Sustainable Development (ISD). They argue that the new roads will inevitably stimulate the use of motor vehicles. There is also a fear that this programme will have a negative impact on valuable nature areas. This fear is not unfounded because new motorways in Warszawa have been proposed without environmental impact analysis. One of the planned routes constituted a direct threat to Otwock Landscape Park (Millard, 1998: 159). Another argument used against the motorway programme is that it is costly and that it will drain resources necessary for investments within the public transport system. Lastly, the Highways Act of 1994 has been criticized for limiting public participation, and not being fully adapted to the requirements of the Statute on Land-Use Planning.[165]

Privatization

Another new issue is environment and privatization. There are major ways of conducting privatization in Poland: (1) capital privatization, under which shares of formerly state-owned companies are sold to third parties, either domestic or foreign; (2) privatization by liquidation, under which the assets of the companies are partly or entirely sold under the responsibility of the regional authorities; (3) mass privatization, under which shares of the enterprises are distributed to the public, with investment funds serving as intermediares (OECD, 1995a). From the environmental protection perspective, it is particularly important to perform environmental audits in the process of ownership transformations. An environmental audit can be used to identify environmental damage and to establish the costs of cleanup and control technology. The audit can provide information that forms the basis for negotiations between the seller and the buyer about the responsibility for these costs (Greenspan-Bell, 1992a: 10095).[166] Environmental audits were used in forty of the 100 first cases

---

[164] In cities such as Katowice and Kraków transport has a larger impact on ambient air quality than industry (personal communications with Wojciech Beblo, 1995 and Krzysztof Göhrlich, 1996).

[165] Personal communication with Andrzej Kassenberg, 1995 and Wojciech Beblo, 1995.

[166] Environmental liability issues are of importance for potential investors in eastern Europe. In a study of 1,000 large western companies potentially interested in investment, it appeared that half of them considered environmental issues (such as liability for inherited contamination) to be a barrier

of capital privatization in Poland. The regulation for this kind of ownership transformations stipulates that the purchaser is liable for environmental damages that have occurred at the site. In contrast, the regulations governing privatizations by liquidation do not regulate environmental issues. Hence, there are no legal provisions requiring environmental audits (OECD, 1995a).[167] It was not until very recently that the Ministry of Ownership Transformation accepted that the environmental obligations for the approximately 400 firms participating in the Mass Privatization Programme should be clearly defined (Millard, 1998).

Municipal waste and consumption

Other issues that have become more emphasized in the 1990s are municipal waste and consumption patterns. According to official statistics, the volume of waste from private households increased by approximately 10 per cent between 1990 and 1995 (GUS, 1997). This figure may, however, be an underestimation. In one press report that was published in 1991 it was stated that the volume of rubbish generated in Poland had increased by 40 per cent since 1989 because of import of foreign products (Manser, 1994: 95-96). From the OECD's environmental performance review on Poland it follows clearly that the capacity to manage the increasing waste streams is limited:

Almost all of the waste goes to landfills without separation (...). Up to 10,000 unauthorized dumps exist, owing to the limited capacity of authorized landfills. The volume of landfills has tripled since 1975. Few municipal dumps are equipped with compactors. For many towns, the amounts disposed are already beyond official capacity. Only 55 per cent of the population is served by waste collection agencies, so proper collection and disposal are difficult. Opposition to new landfills and incinerators has increased, particularly when waste from another voivodship is involved. There is strong concern that local land value will deteriorate even if a state of the art waste disposal facility is installed (OECD, 1995a: 63).

The situation is aggravated by the fact that the recycling schemes for paper, glass, and textiles that were in place during the socialist period have collapsed. The reason for this breakdown is the new price structure for secondary materials (OECD, 1995a: 65).[168]

---

investment as much as legal uncertainties and risks associated with unstable economic reforms (Bluffstone and Panayotou, 1995: 1).

[167] Interestingly, the present approach is not seen as a problem for the Ministry for Environment. In 1995 it stated that "full-scale environmental audits are costly and time-consuming. Smaller firms, which are usually not in very good financial condition, and smaller investors interested in taking them over, cannot afford to finance elaborate environmental audits" (MEPNRaF, 1995: 44).

[168] New problems with municipal waste are also discernible in other parts of the former Soviet bloc. In eastern Germany the amount of garbage collected per inhabitant doubled two years after reunification and soon reached the same level as in the western part of Germany (OECD, 1993). A peculiar phenomenon along the West Bohemian border of the Czech Republic is 'garbage tourism'. Germans from Bavarian and Saxon towns travel to the Czech Republic to deposit rubbish on waste grounds in order to avoid rubbish disposal costs in their own country (Carter, 1998).

5.2.3. Interactions

In the first few years after 1989 the environmental administration was in a clear alliance with the environmental movement. Between 1989 and 1991, a parliamentary commission (belonging under the environmental committee) for contacts with environmental NGOs existed. An office for cooperation with non-governmental organizations was created within the Ministry of Environment in 1991 (BSE, 1995: 120). In addition, the Ministry of Environment was led by persons (Bronisław Kamiński, a former director of the environmental protection department in Kraków voivodship; Maciej Nowicki, a Professor and air protection expert from the Technical University of Warszawa; and Stefan Kozłowski, a geology Professor affiliated with the Polish Academy of Sciences and the PKE) who were all deeply respected by the environmental movement. Policy adoption at the beginning of the transition period was in several instances a collective endeavour of the Environmental Protection Advocacy Coalition. For example, in 1990-91, the Ministry of Environment took the advantage of acting in union with various non-governmental groups in passing the Nature Protection Act and the Statute on the State Inspectorate of Environmental Protection.

However, the EPAC became an increasingly fragile coalition in the new systemic framework. In 1992 the divergence of opinions became more and more manifest and the coalition was dissolved. This process can be explained in two major ways. First, the environmental movement felt uncomfortable with the new ministers of environment, Hortmanowicz and Żelichowski. Under Żelichowski's tenure (1993-97) the conflict between the Ministry's environmental protection mission and its responsibility for the forestry industry was brought to the fore. This was partly related to the fact that the minister had a background in forestry (Millard, 1998). The reason for Hortmanowicz's unpopularity with the environmental movement (and also with the Parliament and various professional bodies)[169] is understandable when the following is taken into account: Hortmanowicz demoted the Chief Nature Conservation Officer and subordinated that post to the new Director General of Forests who issued an order to "reduce the elk population to zero". He also decided to bring forward the deer culling period to a time when mothers would still be suckling their young, and this caused outrage. The Minister was also criticized for using gifts from abroad to serve the aims of his political organization, the Rural Solidarity (Millard, 1998: 148).

Second, and more importantly, consensus began to dissipate about the practical operationalization of the concept of sustainable development. The environmental movement argued that the Ministry of Environment was subordinated to the government's economic policy rather than the environmental policy. The environmentalists began to doubt that the Ministry was truly committed to the goal of sustainable development. The Ministry and the NGO community had diverging beliefs about the proper relationsship between environmental concerns and economic ones in issues such as transport, nature protection, forestry, and waste management. The Ministry was accused of supporting dubious projects (most notably the dam project in Czorsztyn in southern Poland), limiting the influence of the environmental movement within the Ministry, and delaying the elaboration of new environmental laws (Fiałkowski, 1996).[170]

[169] In a letter to the Prime Minister in September 1992 a group of academics accused him of ignorance, incompetence, and breaking the law (Millard, 1998).

The pent-up distrust between the environmental movement and the Ministry of Environment was revealed at a meeting between the two parties arranged by the parliamentary commission for environmental protection in 1995. On this occasion, the Ministry of Environment was represented by Czesław Więckowski, the director of the Department for Ecological Policy which had been established in 1994. He claimed that the environmental NGOs in Poland, as elsewhere in eastern Europe, are divided, weak from an organizational and economic point of view, and focused on criticism rather than on productive work and cooperation (BSE, 1995: 51). In his view, "the non-governmental organizations should not be confined to point fingers at the government, the administration, and the civil servants for their incompetence and lack of action. Instead, the organizations ought to initiate different kind of actions" (*ibid.*: 72). According to Więckowski, it is difficult to identify those environmental organizations that are suitable cooperation partners for the Ministry. Taken together these problems make cooperation between the NGOs and the authorities difficult (*ibid.*: 51).

The representatives of the environmental movement at the meeting declared that there are too few direct consultations between the Ministry and themselves. As Darek Szwed of the Green Federation put it: "the last meeting took place two years ago and after that nothing has happened" (*ibid.*: 95). Dissatisfaction was expressed concerning the fact that no NGO representatives became members of the Inter-ministerial Commission for Sustainable Development. Szwed pointed out that strategic environmental policy documents have been elaborated without input from the NGO community (*ibid.*: 94). Piotr Gliński claimed that the NGO-sector has been marginalized by the Ministry of Environment and other authorities. He concluded that a further marginalization constitutes two threats:

> It may radicalize the environmental organizations, something which already is observable. It may also bring a stop to many public initiatives, especially on the local level. People who are rejected by the administration stop acting or become more radical. I would like the Ministry to take this into consideration (BSE, 1995: 102).

After the meeting in the parliament cooperation between the Ministry of Environment and the environmental NGOs has developed in the following direction. A detailed proposal for cooperation between the Ministry of Environment and the environmental movement was developed by a group of activists chosen through a competition organized by the Ministry. The whole process of developing this proposal was supervised by a steering committee composed of people from both the Ministry and the NGOs. The final version of the proposal was published in February 1997. The document contains a list of changes covering the following issues: (1) increased access to information; (2) the establishment of databases on environmental problems; (3) new procedures for financing NGO activities; (4) suggestions concerning legislative changes regarding public participation; and (5) the establishment of a new organizational unit within the Ministry of Environment responsible for cooperation with environmental organizations.[171]

In the early stage of the transition, the environmental policy-makers met with almost

---

[170] Among some environmentalists, the Ministry began to be referred to as the "Ministry of Environmental Destruction" (Mistewicz, 1997).

[171] Personal communication with Piotr Gliński, 1998.

no opposition from the business community. Bronisław Kamiński[172] recalls that "the legislative changes were carried out in an easy and quick way". The ecological economist Tomasz Żylicz (1991) could write that "as for 'developers', there are hardly any involved in environmental debates in Poland". It was evident that many enterprises had not taken on board the implications of the new situation. The business community had not yet discovered the need to interact with the environmental policy-makers. It is even questionable if it had the capacity to do so because it lacked the expertise, experience, and tools required for performing effective lobby-work.[173]

Industry stayed quiet but there were other groups that were beginning to make their voice heard. When, in 1990, the Ministry of Environment attempted to introduce a 4 per cent fuel charge earmarked for environmental protection, this proposal was strongly opposed by Solidarity. The labour union claimed that, "while the union is for environmental protection, it will not approve any such burden laid on the impoverished society" (Żylicz, 1991: 8). In the second half of 1991, local businesses in northern Warszawa made an informal agreement not to complain to the authorities about the environmental impact of the new trades started by one another (Manser, 1993: 95).

From 1992 onwards the interactions between the environmental subsystem and various economic and industrial subsystems increased substantially. In October 1992 the Council of Ministers reduced the pollution fees by up to 90 per cent (Cole, 1998: 200). This decision was taken after the Ministry of Industry and Trade had convinced the Ministry of Environment that the high fees that had been introduced in 1990-91 were detrimental for industry's economic performance. The government restored the fee level in 1993 (*ibid.*: 201), after protests from enterprises (most notably, power plants) that already had invested heavily in environmental protection measures. In Cole's (*ibid.*: 200) words, "the fee reductions greatly devalued their investments while rewarding enterprises that had done nothing to reduce pollution."

Throughout the 1980s and in the early 1990s the rate of the air pollution fees in the Katowice and Kraków voivodships were double those set by the Ministry of Environment for the rest of the country (World Bank, 1992d). In the autumn of 1992 the Council of Ministers decided to abolish this rule after the Constitutional Court of Justice had stated that double charges in these two provinces were not in accordance with Polish law. The protests against the double fees were initiated by private greenhouse owners in Katowice[174] who argued that the fees constituted an intolerable financial burden for them (Żylicz, 1991).

Alberski (1996: 190) argues that the adoption of new pieces of environmental legislation after 1992 has been delayed partly because of the objections articulated by various interest groups. Millard (1998) claims that the Ministry of Finance has blocked a proposal to introduce economic incentives to increase the use of unleaded petrol. The increased interaction between the subsystems implied that the environmental policy process was beginning to lose its corporatist features and was more and more taking on the properties of a pluralist one.

The environmental subsystem has much more influence on the industrial subsystem

---

[172] Personal communication, 1996.

[173] As Jerzy Śleszyński (personal communication, 1996) puts it: "There was no opposition from the employers because they were not well organized and they did not have enough good experts".

[174] The issue was brought to the Constitutional Court by the Polish Association for Gardeners (Kroner, 1992).

and other economic subsystems in the 1990s than in the preceding decade. Various measures have been taken by the Ministry of Environment to integrate environmental concerns into the sectoral policies:

- The Ministry of Environment cooperates with other ministries in the meetings of the Economic Committee of the Council of Ministers (*Komitet Ekonomiczny Rady Ministrów* or KERM). Another channel is the process of obligatory inter-sectoral consultations for key legal acts and key policy documents. For instance, in 1994 the Ministry of Environment gave its opinion on the new transport and industrial policies and on health policy (Matuszewska and Spyrka, 1995).
- In October 1994 the government established the Inter-ministerial Commission for Sustainable Development. (It functioned until 1997 when it was dissolved and replaced by the Commission for Regional Policy and Sustainable Development.) The main aim of the Commission for Sustainable Development was to "provide a forum for opinion-making and for ensuring congruence between documents, plans, new draft regulations, and the national environmental policy". The commission was made up of high-level representatives of all ministries, central offices, and the environmental protection administration. Its chairman was the Minister of Environment (NFEP, 1997: 43). On the commission's agenda in 1996 were, among other things, the following issues: the strategic development of the country until the year 2010, environmental protection in heavy industry and the energy sector, implementation of Poland's climate policy, environment and health, environment and rural development, financing environmental protection measures, the strategy for environmental education, the national strategy for biodiversity, sustainable consumption patterns, local Agenda 21 development, and environmental protection in the military sector (Komisja do Spraw Ekorozwoju, 1996). A parliamentary group for Agenda 21 has been established (NFEP, 1997: 44).
- In 1992 the Ministry of Environment and the Ministry of Privatization established an inter-ministerial unit for addressing environmental protection issues (such as environmental liability from past pollution) in the privatization process (OECD, 1995a).

To this can be added a number of other initiatives. In 1994 the Ministry of Environment and the Ministry of Industry and Trade initiated cooperation to promote the introduction of cleaner technologies in industry. Until mid-1997 more than 400 companies had implemented cleaner production projects and approximately 1,000 specialists have been trained. The project includes the promotion of environmental management and audit systems. A Responsible Care Programme was adopted by the chemical industry in 1995. It includes measures to limit energy consumption and to modernize environmentally burdensome technologies (NFEP, 1997).

The preparations for EU's internal market increase the cross-sectoral cooperation between the Ministry of Environment and the ministries in charge of economic development (Wajda, 1996a). The debt-for-environment swap arrangement shows that the environmental policy subsystem has had an impact on Poland's international economic relations (see Section 5.3).

5.3. Policy adoption

Environmental policy reform after systemic change has proceeded in three different ways: (1) adoption of new laws; (2) amendment of old laws; and (3) issuance of regulations to old laws (Jendrośka, 1996: 390). In addition, a number of non-legally binding policy documents (White Papers) has been approved and some changes have occurred on the institutional level.

Institutional changes

At the end of 1989 forestry was incorporated into the responsibility of the Ministry of Environment. In 1990 a special commission for Environmental Impact Assessment was established. It is an advisory body to the Minister of Environment consisting of seventy-five members, most of whom are highly qualified experts. The role of the commission is to evaluate EIAs for "especially harmful projects." By so doing the commission assists the Minister in exercising his power to approve permissions concerning these projects (Jendrośka, 1992). In 1991 seven Regional Water Management Boards were created with the task of developing programmes for water use (OECD, 1995a). The principle of basin-based water management was laid down in an amendment to the Water Law in 1997 (MEPNRaF, 1997). (The structure of the environmental protection administration in 1995 is depicted in Appendix 2 and the structure of the Ministry for Environmental Protection, Natural Resources, and Forestry the same year can be found in Appendix 4.)

The National Environmental Policy

Shortly after systemic change, the Ministry of Environment, Natural Resources, and Forestry started to elaborate a new environmental policy. It was to be based on the ecological Round-table in 1989 and adapted to the new economic and political conditions in Poland.[175] After one year of preparations, the Council of Ministers approved the National Environmental Policy on 19 November 1990, and in May 1991 it was adopted by the parliament.[176]

The ultimate goal of the new environmental policy is sustainable development. This concept is defined as "the attainment of a balance between social, economic, technical, and environmental conditions in the process of development" (MEPNRaF, 1991: 3). *All* sectors of the economy should be affected by the policy of sustainable development. The document refers to three factors that may increase the will to implement policies aiming at sustainable development. Firstly, the connection between environment and health "will ensure public support for sustainable development" (*ibid.*). In order to take advantage of this potential, the public needs to be provided with information about the state of the environment and the state of health. Secondly, the document states that policies for a sustainable development can

---

[175] There was no official decision which invalidated the 1988 National Environmental Programme to the year 2010 (personal communication with Waldemar Michna, 1995).

[176] It is not a legally binding document. Its main importance lies in the fact that it indicates the main direction of policy reform.

lead to economic benefits; "environmental protection in its broadest sense will be an ally to a modern, effective, and prudent economy" (*ibid.*). Thirdly, policies for a sustainable development will promote international cooperation, particularly foreign aid: "Poland, because of its catastrophic environmental situation, has an opportunity to attract external financial resources devoted to the reduction of global threats, since they can be more effective in Poland than in developed countries" (*ibid.*).

The document points to eight principles which are of crucial importance for the implementation of sustainable development: control at the source,[177] law-abiding, common good,[178] economization,[179] polluter pays, regionalization,[180] common solution,[181] and staging of long-term plans. The mentioning of these principles reveals that the designers of the new environmental policy were familiar with concepts frequently used within the OECD.

The national environmental policy divides the environmental protection goals into short-term, medium-term, and long-term priorities. The basic goal of the *short-term priorities*[182] is the elimination of such sources of pollution and other environmental abnormalities that constitute an acute threat to public health. These urgent tasks include: (1) the reduction of pollution from eighty industrial plants included in a countrywide list;[183] (2) implementation of a coal quality improvement programme; (3) substantial reduction in air pollution, particularly in Upper Silesia; (4) reduction of deficits in quality drinking water supplies for urban areas; (5) the implementation of an adequate system for the management of industrial and municipal waste; (6) gradual diminution of food crop production on soils affected by toxic substances; (7) reduction of the negative environmental effects caused by transport; and (8) improvement of the environmental situation on state borders.

The *medium-term priorities*, to be completed by the year 2000, aim at limiting the pressure on the environment and achieving conformity with western European standards. Furthermore, the necessity of addressing global environmental problems, such as climate change and the depletion of the ozone layer, is taken into account.

The *long-term priority* of the national environmental policy is "the full introduction of sustainable development principles into the entire economy" (*ibid.*: 12). The cost for implementing the long-term strategy is estimated at $260 billion.

The document anticipates that state intervention will be required not only to complement the market but also to actively promote structural changes.

---

[177] In the following hierarchical order: avoidance of pollution generation, recycling, and neutralization of pollution.

[178] Understood as a greater ecological consciousness and a new ethic behaviour.

[179] "(...) the greatest possible advantage will be taken of market mechanisms, with the necessary maintenance through state intervention" (MEPNRaF, 1991: 4).

[180] Understood as extended influence of the self-governments and regional administrations.

[181] This principle refers to the need to find common solutions for solving European and global environmental problems.

[182] Substantial improvements were envisaged within a few years.

[183] As well as 500 enterprises on lists prepared by the voivodships.

As the economy shifts towards a market system, the activities in environmental protection should be reckoned among branches that require state intervention, thus, which cannot be subjected to the market mechanisms alone. The government is responsible for the elaboration of mechanisms and the scope of such intervention. It is the duty of the government to establish technical standards preventing wasteful utilization of geographic space, raw materials and energy, and to promote technical designs and *structural* changes, which will reduce the negative impacts of natural resources affected by industry and economic development (MEPNRaF, 1991: 9).

The strategy for sustainable development implies, on the one hand, an elimination of the pollution threatening 11 per cent of the Polish area by introducing sustainable patterns of production and consumption. On the other hand, the unspoiled parts of the country, constituting more than one fourth of the area, should be actively protected. Therefore, the document concludes, "a differentiated regional, decentralized approach to environmental protection in our country" is required (*ibid.*: 4). The regional governments are to be given extended rights to pursue regional environmental protection policies, for example, determine standards stricter than the national ones and raising fees and non-compliance fees. The municipalities' position in the process of investment siting procedures is envisaged to be strengthened (*ibid.*: 10).

The group of economic instruments listed in the document include environmental fees, marketable permits, tools which do not strain the state budget (such as tax diversification), and non-compliance fees. The use of economic instruments is justified as follows:

[They] will serve to minimize public costs of environmental protection through an effective differentiation of protection requirements; hence subjects (i.e. polluters), bearing the lowest costs of environmental protection should be faced with the most stringent standards. In practice, this principle will be implemented through the issuance of permits for the use of environmental resources and emission of pollutants. Those permits should be tradeable. Thus, a possessor of a permit should be able to sell that permit to another subject, in accordance with the authority in charge (MEPNRaF, 1991: 16).

Besides the objective of cost minimization, the document identified two additional functions of economic instruments. They should generate revenues for environmental investments and provide motivation for environmental protection. In order to have this effect the rate of the charges must be high enough to provide the polluters with an incentive to abate rather than to pay the fee. According to the document, economic instruments should *supplement* legal and administrative tools.[184]

When the policy-makers designed the new environmental policy they were faced with several strategic choices: should the new policy be something fundamentally new or should it be based on the existing Polish system? How much of the western policy-making experiences should be included? How much innovation was required? Bronisław Kamiński provides the following answer: "Our point of departure was that we should improve the existing environmental management system rather than create a new one. The existing system had many positive features. However, we had to adopt it to the market economy and to include some western experiences".[185]

---

[184] This statement is justified on the grounds that most developed market economy countries do not primarily rely on economic instruments but command-and-control.

[185] Personal communication, 1995.

A comparison between the new national environmental policy of 1991 and the Round-table agreement of 1989 shows a considerable degree of continuity. Since the the Round-table agreement was a reflection of the policy views that were advocated by the EPAC in the second part of the 1980s, it can be concluded that most of the ideas of the 1991 document were conceptualized before systemic change. The only entirely new element in the 1991 document was the appearance of cost-effectiveness as a policy goal which implied that emission trading was recommended as a viable policy instrument.

Some suggestions from the Round-table were absent in the 1991 document. This is related to two factors. Firstly, some of the proposals of 1989 were clearly designed for being implemented in a socialist setting and could therefore no longer be considered. Secondly, some suggestions, such as the idea to set up an *international* environmental bank in Warszawa, were rather unrealistic.

The 1991 document was written in a more structured way compared to earlier policy documents. Details were omitted, priorities were set in a clearer way than before, and more attention was focused on explaining the precise role various policy instruments should play.

Statute on the State Inspectorate for Environmental Protection

The 1991 Statute on the State Inspectorate for Environmental Protection transferred the right to impose fines and to interrupt environmentally harmful activities from the voivodship environmental protection departments to PIOŚ. In this way, PIOŚ was transformed into a powerful enforcement agency, with a legal right to appear as "ecological police". The Statute also provided the basis for a new and comprehensive monitoring system. Fully developed the system will include monitoring for air pollution, surface and ground waters, nature protection, and radioactive contamination. PIOŚ importance increased after it had been given the task of being a watchdog for eighty especially polluting industries. At the start of 1990 the Minister of Environment, Bronisław Kamiński, had published a list of the eighty largest polluters in *Rzeczpospolita*, one of the most important Polish newspapers, in which he called for help from the whole society – local municipalities, environmental organizations, and others – to substantially improve the environmental performance of those enterprises. He declared that "(...) if they do not take effective measures, they will cease to exist" (Waller and Millard, 1992: 175).

Environmental framework law

It is noticeable that the environmental framework law from 1980 has survived the transition despite serious attempts to replace it with a new one. The first attempt was made in the spring of 1991 when a research group on environmental law submitted a new draft law to the Ministry of Environment. Among other things the 1991 draft environmental protection act contained a provision that allowed for emission trading: "the terms of a pollution permit for a plant can be transferred to another plant, either entirely or in part, subject to the approval of the authority who issued the original permit" (Stavins and Żylicz, 1995: 5). The law was approved by the Council of Ministers in the autumn of 1991. However, a group of parliamentarians were dissatisfied with the fact that the new law included both nature protection and pollution control issues. The group managed to eliminate all pollution control provisions and submitted a nature protection bill to the parliament. It was enacted, against the will of the Minister of Environment, in November 1991 (Bowman and Hunter, 1992: 932).[186] The Minister, Maciej Nowicki, wanted to have a comprehensive act covering air, water, waste, and nature.[187] A new law, commissioned by the Ministry of Environment, was drafted in 1996. It was abandoned in 1997, "mainly due to the political dynamics of an election year" (Żylicz, 1998: 13). In August 1998 PHARE gave support for a new attempt to finalize the work on a new framework law (Żylicz, 1998).

Nuclear power

The Polish environmental movement won its hitherto greatest victory in September 1990 when the government decided to abandon the nuclear power plant under construction at Żarnowiec (Waller and Millard, 1992: 177). The decision was taken on economic grounds but the impact of the public opposition against the project was also important.

Harmonization between Polish and EU environmental law

A strategic goal in Poland's foreign policy is the harmonization of national law with directives of the European Union.[188] The association treaty places stress upon cooperation in ecological issues in almost all relations with the EU. Article 71 states that "(...) the policy for the economic and social development of Poland should be directed by the principle of sustainable development. It is essential to guarantee that the requirements of environmental protection be fully included into this policy from the very beginning (...)" (NFEP, 1997: 13). Article 73 of the agreement requires Poland to incorporate environmental considerations into the sectoral policies from the outset (Greenspan Bell, 1992b). Article 80 contains guidelines on issues such as joint actions on environmental monitoring, transboundary and regional air and water pollution, protection of forests, preservation of fauna and flora, and the reduction of greenhouse gases. The parties are encouraged to cooperate through the exchange of information and experts and jointly executed research (Fiedor, 1995: 1). Also issues such as the soil protection, land-use planning, and the environmental impact of

---

[186] Panek (1992) argues that an important reason for the failure to approve the 1991 draft environmental framework act was that it was not sufficiently adjusted to the EU's environmental legislation.

[187] Personal communication with Tomasz Żylicz, 1998.

[188] The EU has hitherto issued more than 200 legal acts pertaining to environmental protection. Of these approximately ninety *directly* relate to environmental protection. The directives concern issues such as air and water protection, waste management, chemical safety, biotechnology, radiation protection, nuclear safety, and nature protection (Równy, 1996). The EU has four legal instruments: ordinances, directives, decisions, and recommendations or opinions. Only ordinances and directives are instruments of transnational law. Ordinances are binding while directives oblige member countries to follow common principles of EU environmental policy and to achieve certain goals within a specified time frame. The way the goals are implemented can be freely chosen. To the legal instruments belong also the judgements of the Court of Justice. It decides, among other things, how the subsidiarity principle should be interpreted (Fiedor, 1995). The implementation of an EU directive is done in two stages. Normally, the EU countries are allowed to wait up to two years before a directive is translated to national law. After that there is sometimes an adjustment period (especially for old plants) from a few years up to ten years (Wajda, 1996d).

agriculture and construction are dealt with here (Równy, 1996).

Between 1990 and 1992 the Institute for Environmental Protection in Warszawa analysed the differences between Polish and EU environmental law. Some forty directives were confronted with the Polish law (Panek, 1992). It appeared that Poland had relatively little regulation compared to the EU within four areas: (1) chemical turnover; (2) genetically modified organisms; (3) public participation in the decision-making process, especially EIA-procedures; and (4) access to environmental information (Wajda, 1996b). Hence, it became clear that important changes in Polish law were needed in these areas.

A number of other differences between EU directives and Polish law have been identified. EU law deals with more narrow and detailed aspects of the environmental protection while Polish law is more general and comprehensive. One striking example is the drinking water directive, which obliges the member states to measure the concentration of various carcinogenic substances that are not measured at all in Poland.

There are substantial differences between Polish and EU legal regulations concerning air pollution from stationary sources. First, Polish air quality standards tend to be more stringent. (However, Polish norms for air pollution from transport are milder than the ones in the EU.) Second, the EU law is based on emission standards grounded on the BAT (Best Available Technology), a principle that does not exist in Poland (Wajda, 1996a). BAT means "the most effective and advanced stage in the development of activities and their methods of operation which indicate the practical suitability of particular techniques for providing in principle the basis for emission limit values designed to prevent and, where that is not practicable, generally to reduce emissions and the impact on the environment as a whole" (Wajda, 1996c).

Poland's compliance status in 1996-98 with respect to its efforts to approximate its environmental legislation to the EU's can be summarized as follows:

Principles and legal institutions. Speaking about principles and legal institutions, the convergence with EU law is high. For example, the environmental framework law from 1980 reflects many of the principles that are contained in the Maastricht Treaty (Article 130r) and the Fifth Environmental Action Programme 1993-2000. The new environmental framework law will fully incorporate EU principles, legal institutions, and definitions (Wajda, 1996b).

The water law. An Executive Order of the Minister of Environment in 1991 deals with classification of waters and conditions to be met by waste water discharges. The Order refers to three related EU directives: 76/464/EEC on pollution caused by certain dangerous substances discharged into the aquatic environment, 86/280/EEC on limit values and quality objectives for discharges of certain substances, and 91/271/EEC concerning municipal waste water treatment. The basin-approach of the 1997 amendments to the Water Law is a leading idea in a draft EU directive on water resources (COM/97/49). In this case it is possible to say that Poland is anticipating the EU (MEPNRaF, 1997: 18).

Air protection. In late 1992 Poland ratified ten automotive rules of the EU related to environmental protection. From 1 July 1995 all imported cars had to be equipped with catalytic converters (Anderson, 1994).

Poland is about to transpose the EU Directive on Integrated Pollution Prevention and Control (IPPC). This directive will oblige Poland to prevent pollution by application of BAT. BAT should serve at defining the permissible emission norms. Emission standards are to be reached by the application of BAT. This change will be carried out in connection with the adoption of a new environmental framework act (Wajda, 1996b).

In an amendment to the 1980 environmental framework law, effective from 1 January 1998, the Council Directive 96/62/EC on assessment and management of emergency levels of ambient concentration of air pollution was transposed to Polish law (MEPNRaF, 1998b).

Waste management. The following pieces of EU legislation have been incorporated in the Waste Law which came into force on 1 January 1998: (1) the Council Directive 75/442/EEC on waste in the scope of enforcing the rules of eliminating and limiting inconvenience and hazards related to waste and the prepapartion of waste management plans; and (2) the Council Directive 91/689 on hazardous waste within the scope of keeping waste inventories, hazardous waste management planning, and regulation of hazardous waste; (3) the Regulation EC/259/93 on supervision and shipment of hazardous waste; and (4) the European Parliament and Council Directive 94/62/EC on packaging and packaging waste. In addition, seven EU legal acts have been transposed through executive orders to the Waste Law (MEPNRaF, 1998b).

Nature protection. The 1991 Nature Protection Act is characterized by a high degree of compatibility with the EU legislation, partly because Polish law and EU legislation have been inspired by the same international conventions on nature conservation (Wajda, 1996b).

Environmental Impact Assessment. Two executive decrees (1995) to the Statute on Land-Use Planning ensure compatibility with EU law (Council Directive 85/337/EEC) on EIA (Wajda, 1996b).

Access to information. The Council Directive 90/313/EEC on the freedom of access to information on the environment was transposed to Polish law in an amendment to the environmental framework law of 29 August 1997 (MEPNRaF, 1998a).

Chemicals and dangerous substances. Full aligment with EU legislation regarding chemical substances will be assured by a new act on chemical substances which is in preparation. This law will deal with, among other things, the following issues: classification and labelling of chemical substances, inventory of hazardous chemical substances, procedures of introduction of new chemical substances to the market, assessment of risk created by new chemical substances, control and prohibition of use and introduction to the market of selected hazardous chemicals, and rules on export of selected hazardous chemicals (MEPNRaF, 1998b).

Genetic engineering. On 27 August 1997 the environmental framework law was amended to take into account Directive 90/220/EEC on deliberate release into the environment of genetically modified organisms (MEPNRaF, 1998b).

Environmental statistics. Poland exchanges information with EUROSTAT, the EU agency for statistics, on environmental issues on a continuous basis. The 1990 agreement between the Polish statistical office, GUS, and EUROSTAT implied that GUS had to refrain from publishing data on the environmental performance of specific enterprises[189] because the EU treats this kind of information as commercial secrets (Forowicz, 1992).

The ongoing harmonization of EU and Polish environmental law is expected to be finalized within five years (Wajda, 1996d). The key event in this process will be the enactment of a new environmental framework law replacing the existing one of 1980. Adoption of the act is expected in the year 2000. The law will

[189] Until 1989 the environmental yearbook published by GUS contained detailed information on the environmental pollution produced by the largest companies.

(...) draw upon the EU legislation philosophy, principles, institutions, and the language, and will ensure public access to information on the environment, in accordance with the Community regulations. The act will take into account such issues as prognostics, environmental impact assessments, protection against pollution, waste management, prevention of major industrial accidents, economic instruments in environmental protection, environmental liability schemes, voluntary participation in environmental protection activities (eco-management, eco-labelling). It will also adopt measures which had previously been absent in the Polish legal system, like application of the best available technologies (BAT/BATNEEC) and differentiation of pollution control requirements with respect to new and existing installations (MEPNRaF, 1998a: 1).

Harmonization between EU environmental law and legislation in Poland will entail substantial economic consequences in that expenditures on environmental protection will have to increase. According to one estimate made by the World Bank, the overall costs for Polish industry to comply with EU standards amount to $21 billion (Business Central Europe, 1997). This is more than ten times more than the annual environmental expenditures in Poland in the mid-1990s.

International cooperation

By mid-1995 Poland had ratified forty-four international conventions and other international agreements related to the environment (BSE, 1995: 120). Among these could be mentioned the UN Framework Convention on Climate Change and the UN Convention on Biological Diversity, ratified by Poland in 1994 and 1995, respectively.[190] Poland is an active member of international organizations such as the UN Environment Programme and the UN Commission for Sustainable Development (NFEP, 1997). Poland is also actively participating in a number of regional environmental programmes. In 1993 Poland, Slovakia, and Ukraine established the Carpathian Euroregion for the promotion of broad cooperation including environmental protection. An International Commission for the Protection of the Odra River was created jointly by Poland, Germany, the Czech Republic, and the European Union in 1996 (NFEP, 1997). The so-called Black Triangle project was launched in the beginning of the 1990s to reduce emission of air pollution in eastern Germany, southwestern Poland, and northern Bohemia in the Czech Republic (OECD, 1995a).

Poland has received environmental assistance from the international community since 1990. Most importantly, the aid has facilitated the transfer of know-how and new environmental protection technologies. With respect to grants, the largest contributors between 1991 and 1996 have been the European Union (via PHARE, its aid programme for central and eastern Europe), Denmark, and the Netherlands (Table 5.2). In addition, the EBRD and the World Bank[191] have offered loans for the

---

[190] Poland was a party or signatory to conventions such as the 1971 Ramsar Convention on wetlands, the 1973 Washington Convention on international trade in endangered species, the 1979 Bonn Convention on the conservation of migratory species of wild animals, the 1979 Geneva Convention on transboundary air pollution, the 1985 Vienna Convention on the protection of the ozone layer, the 1989 Basel Convention on transboundary movements of hazardous waste, the 1991 Espoo Convention on environmental impact assessments, and the 1992 Helsinki Convention on the protection of the Baltic Sea (NFEP, 1997).

[191] In 1990 the World Bank offered Poland a $18 million loan for the improving the effectiveness of environmental policy and improving air quality in Katowice and Kraków (Andersson *et al.*, 1993). The World Bank has also channelled money to Poland via the Global Environmental

improvement of environmental management in Poland (Andersson *et al.*, 1993).

Table 5.2 Environmental aid to Poland, 1991-96 (grants)

	Aid			Aid	
Donor	Million US$	(%)	Donor	Million US$	(%)
Belgium	3.1	0.8	Norway	5.8	1.5
Denmark	57.8	15.2	Sweden	24.2	6.4
EU (PHARE)	133.6	35.1	Switzerland	3.9	1.0
Germany	42.3	11.1	United Kingdom	1.7	0.5
Finland	16.9	4.5	USA	36.6	9.6
Netherlands	49.3	13.0			
Japan	5.0	1.3	Total	380.3	100.0

Source: GUS, 1997. The table does not include environmental loans from the EBRD and the World Bank.

Although foreign asistance has been quite significant, it has never corresponded to more than 5 per cent of the annual environmental expenditures in the 1990s (GUS 1997, OECD 1995a).

Financing

A number of new institutions involved in financing environmental protection measures have emerged in the 1990s:

- The National Fund for Environmental Protection and Water Management is the single largest financing organization for environmental protection in Poland. Its principles and objectives were defined in an amended act of the 1980 environmental framework law, enacted on 1 July 1989. The Fund is an independent organization although controlled by the Ministry of Environment Its main source of income is fees and penalties for the use and pollution of the environment together with repayment of instalments of the loans granted by the fund. NFOŚiGW supports environmental protection projects via (1) soft-loans, (2) grants, (3) subsidies to credits, and (4) equity investments. The most important of these four instruments is soft-loans. When granting soft-loans the Fund applies preferential interest in comparison to the interest set by the National Bank of Poland. The Fund can also offer a grace period for loan redemption and partial remissions. Grants are offered to public utilities for tasks related to monitoring, nature conservation, and environmental education. Via the equity investments, the Fund contributes with capital to enterprises active in the environmental protection sector. In addition, the Fund acts as an administrator of the environmental assistance offered by PHARE of the European Union. A secretariat for "Joint Implementation" of the UN Framework Convention on Climate Change has been set up within NFOŚiGW (NFEPaWM, 1997).
- In 1993 the financing system was decentralized radically when independent environmental funds were set up in each of the forty-nine voivodships and local environmental funds were created in the 2,460 municipalities. The regional funds

Facility. Poland has benefited from one project dealing with biodiversity and a gas to boiler conversion project. The value of these projects is $29.5 million.

operate in a similar manner to the National Fund. The municipal funds are only allowed to provide grants for smaller projects.[192]

- The Ecofund (*Ekofundusz*) was created in 1992 to manage money generated by debt-for-environment swap (see below).
- The Bank for Environmental Protection (*Bank Ochrony Środowiska*) was created in 1990. It is a commercial bank specialized in facilitating environmental protection investments. Some two thirds of its credit portfolio consists of loans provided on preferential terms for environmental projects. The main owner of the bank is the National Fund (45 per cent) while the remaining ownership is held by private domestic capital, the state treasury, and the regional environmental funds as well as a number of companies (Novy, 1996).

The pollution and resource use charges significantly increased on three occasions in the early 1990s: June 1990, December 1990, and December 1991 (Żylicz, 1992d). Most fee rates have increased almost twenty times in real terms in the 1990s compared to the 1980s (Anderson and Fiedor, 1997: 8).

In 1989 the Ministry of Environment decided to peg the fee rates to the official inflation index because of high inflation. In May 1990 the fee rates were decoupled from the inflation index. From that moment the Council of Ministers sets the rates on an annual basis taking into account, among other things, expected inflation (Żylicz, 1992d). In the 1980s environmental investment expenditures had been tax deductible. This privilege was eliminated in 1990. The government still allowed polluters to subtract fee payments from their taxable income (*ibid.*).

Debt-for-environment swap

An important issue on the Polish environmental agenda in the early 1990s was the search for innovative methods of generating financial means for environmental protection investments. One of the ideas that was promoted, especially by the Ministry of Environment but also by environmental organizations such as the Polish Ecological Club, was debt-for-environment swap (or eco-conversion as it is usually called in Poland). The idea gained support from the Ministry of Finance which in 1991 was in the process of negotiating a 50 per cent reduction of the Polish foreign debt to its creditor countries within the so-called Paris Club. (The Paris Club consists of loosely associated creditor countries - mainly within the OECD - which meet on a regular basis for negotiations with their debtor countries.) During the spring of 1991 Poland formed an inter-ministerial committee, chaired by Stefan Kawalec (the deputy Minister of Finance) which elaborated a strategy for debt-for-environment swap with the following key elements. Firstly, the money originating from the swap was to be spent on projects having impact on the global, and to some extent the regional, environmental situation rather than on projects of pure national concern such as noise reduction and municipal waste management. Four priority areas were identified: (1) reduction of transboundary air pollution (sulphur dioxide and nitrogen oxides); (2) reduction of pollution and eutrophication of the Baltic Sea; (3) reduction of emissions of greenhouse gases (mainly carbon dioxide and methane; later protection of the ozone layer was added here); and (4) conservation of the biodiversity (Nowicki, 1993:

[192] In 1995 emission charges were levied on emissions of sixty-two atmospheric pollutants and six types of waste water discharges. The waste disposal charge distinguished 163 substances. In addition, there was a charge for tree removal (Śleszyński, 1996a).

162). Secondly, an independent foundation, the Ecofund, was to be established for the administration and the project management. Poland made clear that it preferred a multilateral rather than a bilateral approach to the debt swaps. Thirdly, the money was going to be spent over eighteen years - in line with the annual debt service profile under the Paris Club - making the environmental swap a long-term commitment. Technically, the debt swap was to be carried out in the following way. Instead of making debt service payments to the creditors, payments would be transferred into an escrow account for each country. These accounts would be drawn upon from the Ecofund for approved environmental projects (Ecofund, 1992: 5). Poland's swap initiative gained momentum in mid-1991 when the United States became the first country to accept a swap making $370 million available for environmental protection. In May 1993 Switzerland agreed to become a donor country to the Ecofund adding another $66 million. In June 1993 France agreed. However, France decided to limit its contribution and converted only 1 per cent of the debt instead of the expected 10 per cent. Still, the French contribution reached $63 million, a considerable amount of money (Andersson, 1996). In 1997 Sweden accepted a 1 per cent swap and in 1998 Italy decided to agree.[193]

The Ecofund is designed to overcome typical problems that appear during the transition period via (1) a long-term financial commitment from western countries and (2) independence from the rest of the environmental protection administration. In this way its vulnerability to economic turbulence and sudden political changes is significantly reduced (Żylicz, 1992b). Furthermore, the Ecofund is capable of attracting the most skilled experts and policy-makers in the field of environmental protection because of the fact that the salaries are compatible with those of the private sector.

5.4. Implementation

At the beginning of 1995 the Ministry of Environment presented a report which analysed the realization of environmental policy in the period 1991-93. The report concluded that for the first time in many decades the state of environment in Poland had improved, especially with respect to air and water pollution. As a result, the Ministry could reduce the number of areas in ecological hazard from twenty-seven to four (Millard, 1998).[194] Indeed, significant environmental improvements have been achieved in Poland after 1989. Table 5.3 shows that the pollution intensity (industrial solid waste, waste water, and air pollution produced per unit of GDP) of the Polish economy was reduced quite radically between 1989 and 1995. Air pollution declined substantially in the period 1989-95. For example, the emissions of sulphur dioxide and particulates decreased by 24 per cent and 44 per cent, respectively (NFEP, 1997: 49). In the same period the generation of hazardous waste was reduced by 15 per cent and there was a 30 per cent decline in untreated waste water discharged to surface waters (*ibid.*: 13). The OECD (1995a: 3) concludes that the reduced pressure on the Polish environment largely is "a result of the contraction of economic activity and the

---

[193] Personal communication with Tomasz Żylicz, 1998.

[194] These areas were the following: Kraków voivodship, Upper Silesia, Legnica-Głogów, and Tarnobrzeg (Millard, 1998).

restructuring of industrial sectors but also as a result of environmental policies adopted and implemented". The main two pillars in the pollution control policy have been (1) the mechanisms for the financing of environmental protection measures (notably the National Fund for Environmental Protection and Water Management) and (2) enforcement of environmental law (executed by the State Inspectorate for Environmental Protection).

Table 5.3 Economic pressure on the environment in Poland

	1989	1992	1995
GDP per capita[a], $1/person	2,300	1,900	2,200
Resource intensity of GDP			
Energy, TOE/$1,000	1.42	1.38	1.20
Energy, TOQE/$1,000[b]	0.89	0.86	0.77
Water, m³/$1,000	171	172	142
Pollution intensity of GDP			
Industrial solid waste, tonnes/$	1.6	1.7	1.4
Waste water, m³/$1,000	50	47	36
Gases[c], *kg/$1,000*	58	43	33
Dust, kg/$1,000	17	9	5

[a] Estimated in 1992 dollars in constant 1992 prices, at the 1992 nominal exchange rate of Polish currency.
[b] TOE = ton of oil equivalent.
TOQE = ton of "oil quality" equivalents: identical with TOE in the case of oil and gas; 2 TOE of coal are assumed to be 1 TOQE; 1 TOE of hydro- and nuclear electricity is assumed to be 3 TOQE.
[c] Excluding carbon dioxide.
Source: Żylicz, 1997.

5.4.1. Financing

Governmental revenues for environmental protection have increased dramatically in the 1990s. The main reason for this was the sharp increase of the environmental charges rates in combination with the high collection rate of these charges. For example, the collection rate was 100 per cent in 1990, 72.4 per cent in 1992, 73.6 per cent in 1994, and 77.0 per cent in 1996 (GUS, 1995 and GUS, 1997). By 1993 the environmental funds were annually spending fifteen to twenty times more in real terms than in 1990 (Wajda, 1993: 19). Expenditures increased from $464 million in 1990 to near $1.5 billion in 1996 (NFEP, 1997: 13). In 1991 the environmental expenditures reached 1 per cent of GDP for the first time in Poland's history (Table 5.4). Thus, Poland's environmental expenditures in percentage terms had become comparable to the level of many OECD countries (OECD, 1995b: 95). If water supply projects are grouped together with environmental investments (which is a common approach in many countries) Poland's expenditures reached more than 2 per cent in 1996 (GUS, 1997).

The National Fund for Environmental Protection and Water Management has experienced a dynamic development in the 1990s. In 1990 its income amounted to $13 million (of which $12 million from fees and fines). In 1994 the annual income had increased to $400 million (of which nearly half originated from fees and fines). Between 1991 and 1996 the fund co-financed 3,700 projects (NFEWaWM, 1996). In the mid-1990s every fourth złoty invested for the environment in Poland originated from the National Fund. Table 5.5 shows that the share of the national, regional, and local funds of the total environmental expenditures in Poland reached 40 per cent between 1991 and 1995 (NFEP, 1997: 13).

By the end of 1994 the Bank for Environmental Protection had a capitalization of more than $200 million. This year 58 per cent of its projects went to water improvement, 37 per cent for air protection, and 5 per cent for soil protection (Novy, 1996). In 1996 the bank approved 3,508 credits (GUS, 1997).

Table 5.4 Expenditures for environmental protection and water management in Poland in the 1990s

Year	Environmental protection (percentage of GDP)	Year	Water management (percentage of GDP)
1990	0.7	1990	0.4
1991	1.0	1991	0.6
1992	1.0	1992	0.6
1993	1.0	1993	0.5
1994	1.0	1994	0.5
1995	1.3	1995	0.3
1996	1.7	1996	0.4

Source: GUS, 1997. The figures for 1996 are preliminary.

Table 5.5 Sources of environmental protection financing in Poland, 1991-95

Source	Percentage
State budget	5
Municipalities	18
Enterprises	32
Environmental funds	40
Foreign aid	5
Total	100

Source: NFEP, 1997: 13.

Between 1993 and 1996 the Ecofund financed 171 projects at a value of $77.5 million. The 1996 expenditures had the following composition: air protection: 21 per cent, protection of the Baltic Sea: 35 per cent, measures against the greenhouse effect: 30 per cent, and biodiversity: 14 per cent. The activities between 1992 and 1996 led to, *inter alia*, reductions of sulphur dioxide (90,000 tonnes per year), dust (183,000 tonnes per year), and chlorofluorocarbons (475 tonnes per year).[195] The Ecofund has participated in the construction or modernization of twenty waste water treatment plants near the Baltic Sea coast. With respect to biodiversity the fund has supported long-term programmes for the protection of the eagle, white stork, black stork, bat, and several other species. The fund has also given support to the construction of waste water plants in protected areas as well as the establishment of twelve buildings for ecological education in the national parks (GUS, 1997).

[195] According to the Ecofund's own estimations, its efforts have resulted in a 20 per cent reduction in the use of CFCs in Poland (personal communication with Maciej Nowicki, 1994).

Major investments

An important part of Poland's environmental protection budget has been devoted to the financing of industrial and municipal waste water treatment plants. Only in 1996 more than 400 such plants were completed. This could be compared with 1980 when fifty-six such plants were taken into operation (GUS, 1997). A large number of new plants are under construction throughout the country. Air protection has become an upgraded policy area in the 1990s, which is reflected in the increase in air protection spending. Poland has been able to finance the installation of advanced technology to reduce air pollution from the largest power plants (Cole, 1995b: 303). In 1996 Poland took into operation technologies capable of eliminating 267,000 tonnes of gaseous pollution which is almost ten times more than was achieved in 1985. The annual installation of waste management equipment was six times higher in 1996 than in 1985 (GUS, 1997) (Table 5.6).

Table 5.6 The capacity of annual installations for environmental protection, 1980-96

Investment	Capacity/ unit	Year					
		1980	1985	1990	1992	1994	1996
Waste water treatment plants	-	56	150	299	341	334	435
Reduction of dust pollution	1,000 t/y	453	190	435.8	129.4	180.5	130.7
Reduction of gaseous pollution	1,000 t/y		27.8	44.5	193.6	77.3	267.5
Treatment of industrial and municipal waste	1,000 t/y		305	604	1167	621	1935
Recultivation of industrial landfills	ha	124	511	346	318	305	754

Source: GUS, 1997.

5.4.2. Enforcement

The environmental administration carried out a strict enforcement policy in the early 1990s. In 1990 approximately 100 heavy industrial polluters were forced to shut down completely or partly (Żylicz, 1997). Twenty-nine enterprises were closed down in 1992 for environmental reasons, 566 legal proceedings were initiated, and 1,357 petitions delivered (EBRD, 1993: 51). Between 1990 and 1993 seven companies on the list of the eighty largest polluters were closed. At another twenty-two plants of this list production was partially halted (OECD, 1995a: 128). Compared to the situation in 1989, the listed enterprises improved their environmental performance as follows: (1) emissions of particulates: 67 per cent reduction; (2) emissions of gases: 44 per cent reduction; (3) waste water discharges: 37 per cent reduction; and (4) waste storage: 42 per cent reduction. In 1994 a number of companies of the "List of 80" were removed and a few new ones were listed (Anderson and Fiedor, 1997).

Fines have been levied for non-compliance with pollution permits. Fines for air pollutants are ten times those of fees while for waste water discharges the rate is progressive. In contrast to environmental fees, fines are not deductible from the tax base and reduce profits directly. PIOŚ provides facilities in non-compliance with a possibility to defer fine payments. If the enterprise has managed to eliminate the problem within five years the payment of the fines is partly or completely cancelled. If, however, obligations are not met the fine is increased by 50 per cent. In mid-1990s some 70 per cent of the polluters in non-compliance had initiated or finalized measures to meet their environmental obligations (OECD, 1995a: 100). In 1992-93 PIOŚ imposed 12,000 fines totalling more than $140 million (Matuszewska and Spyrka, 1995). However, the amount actually paid was several times smaller. As is shown by Table 5.7, the collection rate of environmental fines was as low as 14.8 per cent in 1995 and 33.4 per cent in 1996 (GUS, 1997). A major reason for the low the collection rate of the environmental penalties is that most companies which are obliged to pay fines undertake measures to improve their environmental performance.

Despite the obligation to carry out continuous emissions monitoring, only one in fifty enterprises complied in the early 1990s (EBRD, 1993: 50). An unsolved problem is the untreated saline waste water discharges from the hard coal mines in Upper Silesia. In this case the enforcement mechanisms in the pollution control system have not functioned at all (see Chapter 7). Industrial waste management is another policy area where relatively little progress can be discerned (OECD, 1995a).

Table 5.7 Collection rate of environmental fines

Year	Percentage	Year	Percentage
1985[a]	37.6	1992	22.7
1986[a]	61.8	1993	15.7
1988[a]	9.9	1994	13.2
1989[b]	45.0	1995	14.8
1990[b]	69.2	1996	33.4
1991[b]	41.8		

Source: [a] GUS, 1989, [b] GUS 1991, 1995, 1997.

5.4.3. Environment and development

The environmental performance review on Poland made by the OECD in 1995 stated the following:

By and large, the major ministries have not internalized a commitment to the environment, and existing arrangements have not been effective in holding them accountable for the environmental consequences of their policies: the environment is still seen as an expensive "add-on", which is the responsibility of the Ministry of Environmental Protection, Natural Resources and Forestry (OECD, 1995a: 158).

Furthermore, the OECD noted that "the strategy for implementation of Agenda 21 does not seem to have been approved by the Polish government as a whole" (1995a: 154). Also the Ministry of Environment admits this problem in a progress report about the implementation of Agenda 21 in Poland until 1996: "(...) there remains a widespread feeling that, as elsewhere in the world, the majority of the measures taken are more like declarations, with the existing legal instruments being relatively lenient in their treatment of obligations that the principles of sustainable development be needed" (NFEP, 1997: 45).

Agenda 21 has also been difficult to implement on the local level. According to chapter 28 of Agenda 21, entitled "Local authorities' initiatives in support of Agenda 21", local authorities in each country should achieve a consensus on "a local Agenda 21" for the community by 1996 (United Nations, 1993). In Poland this recommendation has been followed by a fraction of the municipalities. The Ministry of Environment states that only twenty-three municipalities had initiated this process by 1997 (MOŚZNiL, 1997).

A first indication that the integration of environmental concerns into the sectoral policies was going to be problematic was given already *before* the new environmental policy was adopted, when an annex to the document, specifying tasks to be carried out by each ministry, was deleted by the Council of Ministers. Officially it was rejected on political grounds; it was said that it resembled the command-and-control approach of the socialist era. Another reason for deleting this appendix was that the sectoral ministers were not comfortable with specific recommendations.[196]

5.5. Advocacy coalitions and their belief systems

For a rather short episode in the early 1990s environmental policy in Poland was dominated by one coalition, the Environmental Protection Advocacy Coalition (EPAC). It was a strong coalition that contained many of the actors that had been active in environmental policy-making in the late 1980s. The environmental policy reforms of the early 1990s included most of the relevant opinion of the environmental movement of the 1980s. Consequently, there were no clear alternatives to the proposals of the Ministry of Environment. However, in 1992-93 three advocacy coalitions began to be discernible in the environmental policy subsystem: the Mainstream Advocacy Coalition, the Green Advocacy Coalition, and the Efficiency Advocacy Coalition (see Section 5.3 for an explanation of why the EPAC was split).

I. *The Mainstream Advocacy Coalition (MAC)*. This coalition has the following core members: a majority of the employees at the Ministry of Environment and the State Inspectorate for Environmental Protection, most of the environmental funds on central, regional, and local level, the voivodship departments for environmental protection, and some of the older members of the PKE and the LOP. The environmental policy advocated by this coalition could be summarized as pragmatic and incremental.

Deep core beliefs

Man and nature. The MAC does not make any explicit statements on man's relation to nature. But there is an implicit assumption that man should dominate nature.

Approach to the market economy. The coalition has a positive approach to the market economy.

Policy core beliefs

The proper relationship between environment and economy. There should be a balance between goals of the ecological policy and the economic policy, between environmental and economic benefits. Without economic growth, it is argued, there will be no financial resources for environmental protection. There is a strong belief that the companies offering equipment for environmental protection and various

[196] Personal communication with Tomasz Żylicz, 1995.

services have a growth potential which can benefit the Polish economy.

Whose welfare counts? The coalition is an adherent of the concept of sustainable development which implies that the MAC is aware of the need to take into consideration the needs of future generations. However, this issue is not explicitly discussed.

The seriousness of environmental problems. The coalition argues that substantial improvement of the environmental situation has occurred in Poland in the 1990s.

Basic cause of environmental problems. Two major causes for environmental degradation are frequently put forward: lack of money and lack of technology.

Proper scope of governmental vs. market activity. This issue is not addressed by the coalition.

Proper distribution of authority. The coalition is, generally speaking, in favour of a decentralization and regionalization of the environmental management system.

The choice of policy instruments. The coalition discloses an affirmative attitude to existing instruments and approaches. The MAC listens carefully to signals from industry and is sometimes prepared to follow their advice to lessen standards or to reduce the level of environmental charges. The coalition is more and more open for flexible enforcement approaches to lessen the financial burden for the largest polluters. Not surprisingly, it is the MAC that has suggested the introduction of Negotiated Compliance Schedules in the pollution control policy (see Chapter 6 for an explanation of this instrument).

Method of financing. Financing of environmental protection investments is the policy element which is most emphasized by the MAC. The environmental funds should continue to be the backbone of the financing system.

Ability of society to solve the problems. Within the MAC there is a high degree of trust in technological solutions.

Who should participate? Increased public participation is important. However, it is not a goal itself.

Secondary aspects beliefs

Approach to the European Union. The MAC's approach to the EU is the most positive one among the three advocacy coalitions. Virtually no negative consequences of EU membership are identified. Environmental policy-makers within the coalition welcome the move towards BAT due to the approximation to EU legislation. In Sommer's words, "to this time we have regulated the environment taking into account ambient quality standards. It is a noble idea but very unprecise. It is very difficult to regulate".[197]

Priority problems. The problems which are seen as most important for the MAC are air pollution from stationary sources, waste water discharges and waste, both industrial and municipal.

II. *The Green Advocacy Coalition (GAC).* This coalition has the following members: a few scientific authorities,[198] the young generation within the PKE and the LOP, the

---

[197] Personal communication, 1996.

[198] Most notably Stefan Kozłowski, a geology Professor of the Polish Academy of Science who was the Minister of Environment in 1992 and President Lech Wałęsa's chief environmental advisor between 1993 and 1995. In 1993 President Wałęsa established his own environmental council (*Rada Ekologiczna Przy Prezydencie Rzeczpospolitej Polskiej*) which existed until Aleksander

Institute for Sustainable Development, the Green Federation, the Anarchist Federation, various animal rights groups (such as the Workshop for all Beings), various vegetarian communities, various nature protection organizations (for example, Salamander in Poznań), a number of student associations (for example, the Inter-department environmental lobby at Łódź University), the Service Bureau for the Environmental Movement (*Biuro Obsługi Ruchu Ekologicznego* or BORE) and the Green Brigades (*Zielone Brygady*), a Kraków-based magazine for the environmental movement. In addition, there are a number of small green parties in Poland which should be seen as members of the GAC.[199]

Within GAC, two streams can be discerned, one with a fundamentalist orientation and another with a more realistic orientation (compare the situation within *Die Grünen* in Germany). The former are not willing to strike any compromises and frequently resort to extra-legal protests. The latter have a much more pragmatic approach and have the appearance of a green lobby. The difference between the fundamentalists and the realists was obvious in the struggle against the Czorsztyn dam. One group of activists decided to supplement road blocks with more moderate actions. These included: (1) the establishment of a dialogue with the authorities, (2) the creation of a broad coalition of environmental NGOs to coordinate their activities against the dam, and (3) the preparation of an independent scientific report on the dam issue. All of these methods were strongly rejected by the radical wing of the environmental movement (Gliński, 1993: 148).

Deep core beliefs

Man and nature. The actors within the GAC share the fundamental belief that humans are a part of nature and that nature has an intrinsic value.

Approach to the market economy. The GAC supports a development "that goes away from the centrally planned economy but does not get trapped by the market ideology" (Peszko, 1990). Hence, there are strong anti-market sentiments within the coalition. Olaf Swolkień, an environmentalist from Kraków associated with the Green Brigades, writes in his book "A New System – the Same Values" (*Nowy Ustrój – Te Same Wartości*):

> The commercialized media make people all the time believe that in communism a political doctrine ruled and that capitalism (...) represents freedom and the end of history. In reality the most fundamental feature of both systems is that life and politics are subordinated to economic doctrines both of which are materialistic and only based on quantitative criteria (Swolkień, 1995: 45).

Policy core beliefs

The proper relationship between environment and economy. Environmental concerns are implicitly, and sometimes explicitly, considered to be more important than economic ones. Zbigniew Karaczun of the PKE asserts that "the ecological policy

---

Kwaśniewski became the new president in 1996. The council, chaired by Stefan Kozłowski, came to include fifty environmental experts, mostly with a scientific background. The main aim of the council was to promote a transition to sustainable development. In most cases the council's views converged with those of the environmental movement (Rada Ekologiczna, 1994).

[199] Alberski (1996: 177) argues that there were ten ecological parties in Poland in 1993.

should be something like an umbrella, the most important policy".[200]

Whose welfare counts? The coalition emphasizes that the economic, societal, and environmental policies should pay sufficient attention to the needs of the whole of mankind, future generations, and all living species. The philosophy of the GAC is well summarized in an early document produced by the Green Federation. The deep approach to ecology advocated here is understood as "the rejection of the anthropocentric presumption that environment should be protected because its destruction threatens the existence of man" (Peszko, 1990).

The seriousness of environmental problems. Kozłowski (1995: 13) mentions four negative environmental trends which are observable since 1989: a deteriorating health situation in certain areas, increased water deficit, destruction of valuable nature areas in connection with the privatization of land, and increased waste water discharges on the countryside. Other problems that are seen as worrisome by the coalition are the increasing air pollution related to traffic and the increased amounts of municipal waste.

The GAC considers industry's pollution problems to be rather effectively dealt with. Darek Szwed of the Green Federation explains: "Industry is becoming more active. We no longer need to make them aware of the problem. The only problem is the level of production which is determined by the level of consumption".[201]

Basic cause of environmental problems. In the GAC's view, the basic cause of most environmental problems is unrestrained consumption. Swolkień argues that

> (...) from birth to death we are fed with propaganda, whose aim is to create consumers (...) Contemporary advertisement, schools, and media are much more effective than the communist indoctrination. Theoretically, people are free but in order to avoid the pictures, voices, and other impacts of the propaganda and advertisement they would constantly need blindfolds for their eyes and protection for their ears (Swolkień, 1995: 8).

Stefan Kozłowski (1992) is no less critical towards consumption:

> Mindless fascination with the western way of life and economic model may lead to an artificial and unwarranted rise in the consumption of goods that are of little worth and use. Consumption is by nature hostile to the environment and in the long run poses a threat to the human civilization as a whole.

It is mainly due to the emergence of new consumption patterns that the transition has produced new threats to the environment (Kozłowski, 1995: 13).

Proper scope of governmental vs. market activity. The government should restrain the market forces so that the overall goal of sustainable development could be achieved.

Proper distribution of authority. The GAC has a clear-cut "small is beautiful approach" to environmental issues.

The choice of policy instruments. There is a strong conviction that a green tax reform (which implies higher taxes on pollution and the use of natural resources and lower tax on labour) should be carried out (Szwed, 1996). The GAC does not consider GDP to be a good way of measuring the wealth of a country because certain values

---

[200] Personal communication, 1996.

[201] Personal communication, 1996.

cannot be translated into the language of money. Moreover, economic growth is sometimes achieved at the cost of the environment. Instead of GDP the coalition proposes the introduction of alternative ways of measuring wealth such as the Human Development Indicators (HDI) (Maciejewski, 1994).

Method of financing. The coalition has no objections to the financing system which has emerged in Poland.

Ability of society to solve the problems. There is not much trust in technology or large-scale projects of any kind. End-of-pipe solutions are firmly rejected.

Who should participate? The members of the GAC are strong advocates of decentralization and direct democracy (Gliński, 1994). Public participation in the decision-making process has an intrinsic value and enhances the quality of public policy. There is a widespread belief that the realization of sustainable development requires a radical change in society's ecological awareness. For this purpose a far-reaching educational reform is necessary (Kozłowski, 1995: 20).

Secondary aspects beliefs

Approach to the European Union. The coalition has a critical approach to the EU from an environmental viewpoint. It is generally believed that the EU member countries have not achieved a sustainable development and thus cannot serve as models for Poland in this respect. Many members of the GAC are against the EU because of its promotion of free trade. Free trade, it is argued, tends to lead to over-consumption and various environmental problems. In commenting on Poland's strategy for integration with the EU, the Institute for Sustainable Development identifies a number of weak points. First, the economic threats related to the integration have been too narrowly defined. "Environmental costs have been disregarded (e.g. increased gas emission, growing demand of energy from conventional sources, threatened biodiversity) and these costs may emerge as a by-product of the much expected and extensively covered economic acceleration" (ISD, 1998: 5). Second, the institute is critical of the fact that environmental tax reform is not referred to as a means of combating unemployment. Third, there is no general assessment of the consequences for the environment in Poland: "Environmental protection in the integration strategy should reflect problems and issues which will or should occur after strategic assessment of selected sectors has been conducted in transport, agriculture, and fishing. The lack of such assessments prevents us from defining desirable environmental impacts of the integration process" (*ibid.*: 6-7). The threat to Polish agriculture is a special concern for the ISD. The institute argues that "an indiscriminate transfer of western farming patterns may lead to a loss of the attributes of rural areas which would otherwise support organic farming, biological diversity, and recreation/tourism" (*ibid.*: 8).

The institute argues that the challenges facing Poland in connection with the integration with the EU are of such a dimension that the national environmental policy of 1991 must be replaced by a new policy document adjusted to the new economic and political situation.

Priority problems. The GAC considers the following issues to be the most important ones: consumption, municipal waste, transport, nature protection, and animal rights.

III. *The Efficiency Advocacy Coalition (EAC).* The EAC is the smallest advocacy coalition in the environmental policy subsystem. Its founding father is Tomasz Żylicz, a Professor at the Warsaw Ecological Economics Centre of the Faculty of Economics at the University of Warszawa. Based on his acquaintance with US environmental policy experience,[202] he claimed as early as the 1980s – before any other policy-maker, researcher or environmentalist – that cost-effectiveness should be a major criterion for a successful environmental policy. During the course of the 1990s this viewpoint has gained more and more currency. From 1992-93 a "constituency for efficiency" or the Efficiency Advocacy Coalition did appear in the environmental policy subsystem[203] with the following key members: the Warsaw Ecological Economics Centre at the Warsaw University, the Polish chapter of the European Association of Environmental and Resource Economists (EAERE),[204] the Ecofund, certain officials within the environmental administration (especially some voivodships and municipalities in southern Poland), a few officials at the Ministry of Environmental Protection, Natural Resources, and Forestry and a few members of the Sejm Committee for Environmental Protection, Natural Resources, and Forestry. Furthermore, the Polish power industry also belongs to this coalition because of its advocacy of emission trading. The power sector has realized that it could substantially cut the costs for complying with its air protection obligations if this instrument was used in Poland.

Deep core beliefs

Man and nature. The EAC does not make any explicit statements on man's relation to nature. But it is implicitly assumed that man dominates nature.

Approach to the market economy. The EAC is a strong advocate of the market economy.

Policy core beliefs

The proper relationship between environment and economy. In the coalition's view, if appropriate policies were used, environmental and economic goals would not be in

---

[202] Żylicz was a guest researcher at the University of Wisconsin-Madison in 1977-78 and at the University of Colorado in Boulder in 1988-89.

[203] In establishing the EAC, Tomasz Żylicz acted as a policy entrepreneur. Such an actor tends to use the following strategies (Mintrom and Vergari, 1996: 424): (1) "Policy entrepreneurs seek to sell their policy ideas and, in so doing, to promote dynamic change. Contributors to the agenda-setting literature suggest that policy entrepreneurs use several activities to promote their ideas. These include identifying problems, shaping the terms of policy debates, networking in policy circles, and building coalitions" (*ibid.*: 423). (2) Policy entrepreneurs develop strategies for presenting their ideas to others (*ibid.*). (3) Policy entrepreneurs must have access to resources, either organizational or personal. Policy entrepreneurs based in universities or "think tanks" may use their organizations as "safety havens" for exercising intellectual freedom (*ibid.*). (4) Policy entrepreneurs must have personal resources that include intellectual ability, knowledge of policy matters, leadership, reputation and contacts, strategic ability, and tenacity (*ibid.*: 424). All of these points shed light on the way Tomasz Żylicz, who has a reputation of being one of Poland's most knowledgeable and respected ecological economists, has formed the EAC. He has established a wide network including researchers, policy-makers, politicians, industry representatives, journalists, and NGO representatives with whom he is in frequent contact. A central part of Żylicz's strategy to create a constituency for efficiency has been to arrange a demonstration project of emission trading and to translate western literature on ecological economics (see Chapter 6). Warsaw Ecological Economics Centre, led by Żylicz himself, and the Ecofund, where he is a member of the board, are units that have served as think tanks for the EAC.

[204] The Polish chapter of EAERE was formed in June 1990.

conflict with each other. If environmental policy became more integrated into the economic policy, conflicts between economic and environmental goals would be reduced. A better design of environmental policy would hence reduce the conflicts between environmental and developmental goals. A major problem is that policies tend to be adopted based on imperfect understanding of costs and benefits (Żylicz, 1997: 11). The most useful role played by economic instruments is that they help to minimize overall costs of environmental protection "through an efficient and non-arbitrary differentiation of control requirements" (*ibid.*: 5). "It is (...) not the lack of money but rather the inability to spend the money cost-effectively which postpones recovery and makes environmental policy difficult to enforce" (*ibid.*: 7).

Whose welfare counts? Within the coalition there is a conviction that there should be no privileges or exemptions for any sectors.

The seriousness of environmental problems. The environmental situation in Poland is improving but there are problems which are not sufficiently addressed, such as air pollution from mobile sources and from low stacks.

Basic cause of environmental problems. An important cause of environmental problems is poorly designed environmental and economic policies.

Proper scope of governmental vs. market activity. The EAC is the most market-oriented advocacy coalition in the environmental policy subsystem. Still, it considers a well-functioning state to be a condition for the market.

Proper distribution of authority. There is widespread support for a decentralization and regionalization of the environmental management system.

The choice of policy instruments. Stress is laid upon using the existing resources in an optimal way by harnessing market-based policy instruments such as marketable permits. The EAC advocates the use of marketable permits for, for example, improving air quality (see Chapter 6 for more on this), eliminating the use of CFCs and other ozone-depleting substances, and protecting natural resources from overexploitation (Żylicz, 1992c).

Method of financing. The coalition considers the environmental fund system to function relatively well. Still, the disbursement mechanisms of the environmental funds (the Ecofund excluded) are not sufficiently transparent (Żylicz, 1997: 14). In the long run, the coalition would like the role of the environmental funds to be decreased. The future financing system should be based on the Polluter-Pays-Principle.

Ability of society to solve the problems. In the EAC's view, a more efficient use of existing resources increases the ability to address environmental problems. It is underscored that the environmental performance of industry can and should improve. This should be done with so-called win-win investments,[205] well-targeted end-of-pipe investments,[206] and the introduction of clean technology.

Who should participate? The coalition is in favour of increased public participation in the policy-making process.

---

[205] Personal communication with Bogusław Fiedor, 1995.

[206] Personal communication with Jerzy Kwiatkowski, 1995.

Secondary aspects beliefs

Approach to the European Union. Criticism is directed towards the EU because of its reluctance to rely on market-based policy instruments. As Grzegorz Peszko[207] puts it: "EU environmental policy does not care about cost-effectiveness. EU environmental policy does not care about flexibility, which does not open the door for too much discretion". In a similar vein, Jerzy Kwiatkowski, a former employee of the Ministry of Environment,[208] claims that "Poland is still far away from a situation where we will be able to waste money in the pattern western countries did because the western countries began to tackle their environmental problems not when the problems appeared but only when the money was there".

Representatives of the EAC criticize the Ministry for Environment for its approach to environmental investments. Its implementation plans for environmental policy "consist of lists of investments projects without any sense of prioritizing, not to mention the appropriateness of calculating the costs of investment. All the policy-makers have done is to list a shopping-list of investments and to add their costs. No policies, no targets, no instruments, just investments".[209]

The central elements of the advocacy coalitions' belief systems are depicted in Table 5.8. A comparison between the three belief systems shows that the MAC and the EAC have similar deep core beliefs about man's proper relationship with nature and the approach to the market economy. The GAC has different views on both of these matters.

Policy-oriented learning

Policy-oriented learning in the 1980s facilitated environmental policy reform in 1989-91 and gave the environmental policy-makers a "flying start" in the new systemic framework. One obvious example is the debt-for-environment swap arrangement with the Paris Club in 1991. The Ministries of Industry, Agriculture, Health, Construction, Privatization, and Transport all wanted to have a swap but the Ministry of Environment was selected by the Ministry of Finance because it was the first Ministry to elaborate a concept paper and it was recognized as being the Ministry best prepared to do such a thing (Żylicz, 1992a). Another example is that the 1990 National Environmental Policy was adopted by the parliament before any sectoral policies were approved. These events indicate that the learning process was more dynamic in the environmental subsystem than in most other subsystems.

Environmental policy reform in 1989-91 was further facilitated by the active role played by a few enlightened policy-makers at the Ministry of Environment.

It is rather difficult to categorize the approximation of Polish law to EU directives in terms of learning. Thus far, the harmonization process has been a matter of copying and policy-taking rather than learning and policy-making. Still, there is potential for policy-oriented learning, which is related to the continuous information exchange and the necessity of bureaucratic adjustments and political reorganizations (EBRD, 1993).

[207] Personal communication with Grzegorz Peszko, 1995.

[208] Personal communication, 1995.

[209] Personal communication with Grzegorz Peszko, 1995.

Table 5.8 Advocacy coalitions and their belief systems in the 1990s

	Mainstream AC	Green AC	Efficiency AC
Main members			
	The Ministry of Environment, the National Fund, the voivodships environmental funds, PIOŚ, voivodships' environmental departments.	The environmental movement, most important: PKE, Green Federation, Institute for Sustainable Development, and the Green Brigades.	Ecological economists in EAERE, Warsaw Ecological Economics Centre, the power sector, and the Ecofund. A few officials within the public administration.
Deep core			
	Man dominates nature. Positive approach to the market economy.	Man part of nature. Sceptical/negative approach to the market economy.	Man dominates nature. Positive approach to the market economy.
Policy core			
Environment vs. economy	Formal commitment to sustainable development. Trade-offs between environment and economy.	Environment most important.	There is no true contradiction between the two goals.
Whose welfare counts?	No clear-cut preferences but sectoral privileges not excluded.	The whole of mankind, all animals, and future generations.	No sectors or groups should be given special advantages.
The seriousness of the problems	The environmental situation is improving.	Industry improves but many other things go in the wrong direction.	Industry can and should improve its environmental performance.
Basic cause of the problems	Lack of money and technology.	Consumption.	Poorly designed economic and environmental policies.
Market vs. the state	No clear-cut preferences.	The state should restrict the market forces.	State regulation is a prerequisite for a functioning market.
Distribution of authority	Decentralization.	Decentralization.	Decentralization.
Choice of policy instrument	Present tools + negotiated compliance schedules	Present tools + green tax reform.	Present tools + emission trading.
Method of financing	Environmental funds.	Environmental funds.	Environmental funds. In the long run: PPP.
Ability of society to solve the problems	Technological optimism.	Technological pessimism.	High ability if resources are used in the right way.
Who should participate?	Public participation important.	Public participation is crucial.	Public participation is important.
Secondary aspects			
Priority problem(s)	Air protection, waste water discharges, waste.	Municipal waste, transport, nature protection, animal rights.	Public health (e.g. related to ambient air quality).
Approach to the EU	Positive.	Negative.	Critical acceptance.

There are very few clear-cut indications of policy-oriented learning *across* belief systems. One example of such learning is that the MAC, on the EAC's recommendation, has begun to reform the National Fund for Environmental Protection and Water Management to spend its money in a more cost-effective way. Another example is that the realists within GAC have become advocates of emission trading as a result of interaction with the EAC. This interaction is facilitated by the fact that some of the key members of the EAC – such as Wojciech Beblo, Grzegorz Peszko, and Tomasz Żylicz – were leading figures of the environmental movement in the 1980s.

As regards policy-oriented learning within coalitions the following observation could be made. The EAC has learnt that a "constituency for efficiency" must be *created*. For this purpose a pilot project was arranged in Chorzów and two major studies on sectoral and regional emission trading have been carried on (see Chapter 6 for more details). Furthermore a number of western books and reports on economic instruments have been translated into Polish. The 1989 OECD report *Economic Instruments for Environmental Protection* was published in 1990. At the beginning of 1992 *Project 88* – the US Environmental Protection Agency's report on economic instruments – was published and widely distributed (Żylicz, 1992d). In 1996 the book *Principles of Environmental and Resource Economics*, edited by three prominent western economists, appeared in Polish.

The GAC has learnt that its policy proposals must be well prepared. It has become obvious to a large segment of the NGO community that it is not enough just to protest. Kassenberg explains:

In the 1980s it was much easier because all you needed to say was that the communists are stupid, that they had done the wrong things. That was your argument. It was not necessary to prove that we were right. Now you must work really hard to prepare an argument. If we want to blame the government for not taking into account environmental issues in the privatization process, we must prepare a report within two months.[210]

The agenda of the GAC is partly imported from western NGOs. For instance, the Polish green movement has learnt about ecological tax-reform and integrated this approach in its strategy for realizing sustainable development. As one NGO representative puts it, "there are a lot of new issues because we are open to the influence of western ideas".[211] The MAC has learnt from western counterparts that compliance with environmental standards can be achieved by negotiations. Industry has learnt to get better integrated with the environmental policy subsystem because it has realized that environmental policy has an impact on the economic performance of the companies.

In the 1990s the role of the Polish Academy of Sciences has diminished. It has lost political status, economic support from the state has been reduced, and there is a brain drain in the science sector.[212] There are no natural fora enhancing the process of

[210] Personal communication, 1995.

[211] Personal communication with Piotr Gliński, 1995.

[212] Total expenditure on science in Poland was approximately 0.8 per cent of GNP in 1997, half as much as the average for countries in the European Union. The Ministry of Environment claims that "it is possible to notice some symptoms of a research and development crisis in Poland" (NFEP, 1997: 123). This crisis is revealed in reduced employment in the scientific sector, the disappearance of

policy-oriented learning but to a certain extent the Sejm's commission for environmental protection fulfils the role of a forum (although not based on professional standards) for policy-oriented learning. Jan Komornicki, the chairman of the Sejm Committee for Environmental Protection, Natural Resources, and Forestry between 1993 and 1997 claims that

> In the beginning of the tenure, the members of our committee agreed that we should deal with environmental policy without involving party politics. Hence, we try to base our work on content-related and professional reasoning. Usually we are able to reach decisions to which all members can agree. In general there is consensus.[213]

Jan Komornicki himself has played the role of a policy broker in the committee.

5.6. Summary and conclusions about policy change

The system of financing, enforcement, permitting, and standards that was developed in the 1980s has been strengthened in the 1990s. Despite economic regression in the early 1990s, a number of environmental policy reforms were carried out and several new initiatives were launched. To these belong the sharp increase in the rate for environmental fees and fines and the debt-for-environment swap agreements with some of Poland's western creditor countries.

The national environmental policy of 1991 is the main environmental policy document of the 1990s. The document emphasized the concept of sustainable development and advocated the introduction of emission trading in the pollution control system.

Poland's environmental performance has improved substantially in most sectors since 1989. This is the result of environmental policy reform, contraction of polluting activities, and economic reform.

The main driving force for environmental policy reform in the 1990s has been Poland's commitment to becoming a member of the European Union. This is a strategic political and economic goal for virtually all major political groups in Poland. Poland's association agreement with the EU has implied that environmental protection has been linked to the foreign policy agenda. In Gawlik's words, "the fact that all legislation is analysed with a view to incorporation into the EU's legislation gives the ecologists power in the government".[214] Important international environmental events, such as the UN Conference on Environment and Development in Rio de Janeiro in 1992, have not been directly linked to the foreign policy agenda and have therefore been of much less importance.

In the dominating advocacy coalition of the 1980s, the EPAC, everyone had a general and rather vague commitment to environmental protection. Systemic change implied that the actors in EPAC began to look at environmental problems and their solutions in different ways. Rather soon after systemic change the EPAC was split

---

highly qualified young staff, inadequate equipment, and inefficient cooperation between scientific centres and enterprises (*ibid.*: 122-123).

[213] Personal communication, 1997.

[214] Personal communication with Radosław Gawlik, 1995.

into three coalitions: the Green Advocacy Coalition, which considers environmental protection to be the most important public policy goal; the Efficiency Advocacy Coalition, calling for efficiency-oriented approaches; and the Mainstream Advocacy Coalition, emphasizing the role of financing and flexible enforcement. In the early 1990s industry began to interact with environmental policy-makers and the policy process more and more took on the properties of a pluralist one.

PART III

SECTORAL POLICIES

6. AIR POLLUTION FROM STATIONARY SOURCES

This case study analyses Poland's policy of reducing air pollution from stationary sources. This issue is interesting to study from several respects. It represents a policy area where substantial improvements with respect to policy impacts can be noted in the 1990s as compared to the 1980s. Second, attempts have been made to achieve a conceptual change, *viz.* the introduction of emission trading, in the instrumentation for dealing with air pollution from stationary sources. Third, the issue is well-documented and has been on the environmental agenda in Poland long enough to qualify for an analysis from an ACF perspective.

This chapter has a somewhat deviating structure as compared to Chapters 3, 4, and 5. First, the reader will find information about policy adoption and interactions between the actors in the same section. Second, the chapter is dealing with much more narrow and technical aspects of environmental policy. Third, the case study does not deal with the policy core but the secondary aspects of the advocacy coalitions' belief systems. The reason for this is that this case study only deals with a part of the environmental policy subsystem. According to the ACF, beliefs that are not pertainable to the whole subsystem belong to the category of secondary aspects beliefs (Sabatier, 1998).

Section 6.1 provides a general introduction to the problem of air pollution, especially the problem of acid rain. Section 6.2 looks at the policy processes and policy adoption. Section 6.3 focuses attention on four major issues in the air protection policy: standards and permits, emission trading, enforcement, and fees and financing. Section 6.4 deals with implementation. Section 6.5 deals with the advocacy coalitions and their beliefs systems and addresses policy-oriented learning. Section 6.6 draws conclusions about policy change and summarizes the chapter.

6.1. Introduction

In a famous Polish legend, sulphur mined near Kraków saved its residents from being devoured by a dragon. The myth says that the dragon was killed by a shoemaker boy who killed the beast by placing tar and sulphur inside a sheepskin that subsequently was eaten by the dragon (Manser, 1994: 1-2). In modern times the residents in Kraków have learned to fear sulphur more than dragons. During the past few decades they have witnessed a process of destruction of the historical monuments in Kraków's unique historic centre, which has been placed on UNESCO's list of World Cultural Heritage Monuments (World Bank, 1989d: 13). The problem of sulphur dioxide is known to all Polish towns that use coal for heating. Air pollution has been high on the environmental agenda since the 1980s because of its health impact and damage to ecosystems. Poland's air pollution emissions have also been a matter of international concern because of the export of pollutants, notably sulphur dioxide, to other European countries.

Air pollution – a background

Although air pollution is a relatively new public policy issue, the phenomenon has appeared ever since the industrial revolution.[215] In the nineteenth century air pollution from furnaces, smelters, and chemical works gave rise to the first efforts to address the problem. Following the 1952 London smog, the United Kingdom adopted the Clean Air Act. In the 1950s and 1960s similar legislation appeared in many other countries in western Europe, North America, and in Japan (Tolba *et al.*, 1992: 6).

One of the most commonly appearing air pollutants is sulphur dioxide. Most sulphur emissions from human activities originate from the burning of coal and petroleum products, petroleum refining, and nonferrous smelting (Wijetilleke and Karunaratne, 1995: 43). Sulphur dioxide emissions have two dimensions. The first dimension is the impact on public health at the local level.[216] The second dimension is the transboundary effects that sulphur dioxide causes because of acidification. In the 1960s, Svante Odén, a soil scientist at the agricultural university in Uppsala, Sweden, demonstrated a link between industrial emissions in various European countries and acidification damage in Sweden (McCormick, 1992: 183). It subsequently became widely accepted that acidification may occur thousands of kilometres from the pollution sources. Acid deposition poses a threat to lakes and their aquatic life, forestry, agriculture, and wildlife (*ibid.*: 51).

6.2. Policy adoption and the policy process

Before 1980

Air pollution was not unknown to the socialist government that was installed in Poland after the second world war. In 1954 a decree about the Sanitary Epidemiological Service (SANEPID) was issued. The decree gave SANEPID the authority to control enforcement of sanitary regulations and guidelines with regard to human health and life. At this stage SANEPID focused exclusively on dust emissions. In 1964 the Social Foundation for Revaluation of the Monuments in Kraków (*Społeczny Fundusz Rewaloryzacji Zabytków Krakowa*) was created by the voivodship authorities to restore buildings of historical value. It was soon clear that the revaluation activities could not keep pace with the destruction processes.[217]

Standards for permissible concentrations of air pollutants were introduced by the Council of Ministers in 1965 (Karaczun, 1993: 25). Another important step was taken in 1966, when the Parliament passed Poland's first law on air protection (Karaczun,

[215] It should be noted that air pollution is not merely the amount of pollutants released into the atmosphere. As Wijetilleke and Karunaratne (1995: 16) have pointed out, "topography, weather conditions, time of day, and the kinds of pollutants, and the interactions among them help all to determine pollution levels".

[216] At least one billion people worldwide are considered to be exposed to unhealthy levels of atmospheric sulphur dioxide (Wijetilleke and Karunaratne, 1995: 46). It is known that children and both healthy and at-risk adults are vulnerable to the effects of sulphur dioxide. Individuals who suffer from chronic respiratory diseases may experience coughing and breathing difficulties when ambient concentrations of this substance increase (*ibid.*: 5).

[217] Personal communication with Jerzy Wertz, 1997.

1993: 25). *Trybuna Ludu*, the communist party's main newspaper, was soon to call it "the best formulated and most progressive law of this sort in the world" (Cole, 1995a: 314). Among other things this law introduced a fine for excessive emissions of certain air pollutants (Radecki, 1992).

In the 1970s measures against air pollution focused on particulates. Several companies specializing in dust reduction control were established. The power sector's impact on air quality was discussed among air pollution experts affiliated with the Polish Academy of Sciences. To improve air quality they suggested the use of coal of higher quality in inversion. However, this idea was not implemented.[218]

Poland chose to base its air protection control system on national ambient air quality standards. This was related to the fact that environmental policy in Poland emerged in the 1960s and 1970s mainly as a result of efforts made by natural scientists while environmental protection still was of little interest for lawyers, economists, and technicians. The Polish approach automatically came to reflect the bias of the professional group that was most active at the time. Also the political authorities liked the idea of ambient air quality standards.[219]

The 1980s

In the 1950s, 1960s, and 1970s the air protection policy had been shaped by a narrow group of experts. In the 1980s the policy-making process became more dynamic by the involvement of new actors such as environmental organizations and the Ministry of Environmental Protection and Natural Resources. By and large, the policy process was identical to the one that was described in Chapter 4 on national environmental policy in the 1980s.

During the Solidarity period in 1980-81, articles about the dramatic air pollution situation in southern Poland appeared for the first time in the daily press. Journalists highlighted issues such as the destruction of the old town in Kraków, the environmental impact of the aluminium smelter in Skawina, and the lead pollution of children in Upper Silesia.[220] Throughout the 1980s, scientific evidence on the damage caused by air pollution – such as forest dieback, various illnesses, economic losses, and destruction of historical monuments – accumulated. In the official statistics of 1982 it was stated that 6.4 per cent of the country's forests showed damage due to air pollution, of which 2 per cent manifested "severe" damage.[221]

A number of studies about the economic losses caused by air pollution were made in the 1980s. According to one of those studies, the value of damage to forests as a result of air pollution was estimated to be 50 billion old zloties per year, or more than twice the 1986 expenditures for air pollution control. The costs of deacidification of soils acidified by air pollutants were calculated at 45 billion zloties and the damage caused by air pollution to historic structures and monuments and pieces of art was estimated

---

[218] Personal communication with Włodzimierz Bojarski, 1995.

[219] Personal communication with Jerzy Sommer, 1996.

[220] Personal communication with Krystyna Forowicz, 1995.

[221] In the mid-1980s it appeared that the situation was even worse. It was shown that over half of Poland's territory contained a concentration of sulphur dioxide, 20 micrograms per cubic metre, at which coniferous forests begin to show damage (Kozłowski, 1986). It was revealed that the forests of *Góry Izerskie* and *Sudety* were among the most threatened forests in Europe.

at annual losses of 75 billion zloties (World Bank, 1989d).

Another characteristic feature of the 1980s was that more and more attention was focused on the most polluted area, Katowice voivodship. It was revealed that this region, with 2.1 per cent of Poland's territory, generated a quarter of the country's gaseous and dust emissions. Urban life expectancy in Katowice was the lowest in Poland and in certain parts of the voivodship, such as Bytom, average infant mortality was close to 30 per 1,000 compared to the national average which was 18.5. At the end of the 1980s it was no longer a secret that the Katovice voivodship suffered from the highest incidence of premature births (8.5 per cent), genetic birth defects (10.1 per cent), and spontaneous miscarriages (Manser, 1994: 21-25).

Air pollution was recognized as a developmental barrier for the Katowice voivodship in the mid-1980s. The regional authorities elaborated a long-term environmental protection programme that recognized air pollution as the most important environmental problem for the region. It called for an increased use of economic instruments, strict enforcement (including closures), and integration of environmental concerns into the industrial development of the voivodship (Wojewoda Katowicki, 1985).

Monitoring of air pollution, mainly executed by SANEPID and OBiKŚ in the voivodships, improved year by year in the 1980s. Most progress was made in Kraków, a city that introduced, as the first city in Poland and eastern Europe, an automatic monitoring system for air pollution (MONAT) in 1984. In order to optimize the air protection strategy, this system was used by the authorities to identify the main pollution sources affecting ambient air quality.[222] Monitoring also improved in industry; by the mid-1980s there were roughly 500 air monitoring stations in operation in various production units (World Bank, 1989f: 41).

The creation of the Ministry of Environmental Protection and Natural Resources in 1985 immediately entailed more focus on air protection. One explanation may be that the first Minister of Environment, Stefan Jarzębski, was an air protection specialist himself.

Poland ratified the Convention on Long-Range Transboundary Air Pollution (LRTAP) in 1985. As the only eastern bloc country Poland refused to sign the 1985 sulphur protocol[223] of the LRTAP (Haas *et al.*, 1993). The Ministry of Environment claimed that 30 per cent reductions were unrealistic and that Poland did not want to make a promise it could not keep.

Poland's position *vis-à-vis* the sulphur protocol implied that the country appeared as a laggard state in European environmental politics. In the second part of the 1980s various actors in the West - such as governmental agencies, environmental organizations, and scientific organizations – took up cooperation with their Polish counterparts to, *inter alia*, increase their awareness of the acid rain issue. Bilateral cooperation emerged between Poland and countries such as Sweden, Germany, the USA, and the Netherlands. More and more frequent contacts were developed between Polish environmental NGOs (most notably, the PKE) and western environmental organizations.[224]

---

[222] Personal communication with Bronisław Kamiński, 1996.

[223] Those countries who signed this protocol committed themselves to reduce their sulphur emissions by 30 per cent in the year 1993 compared to the 1980 emission level.

[224] Personal communication with Józef Kozioł, 1995.

A milestone in Poland's air protection policy in the 1980s was the establishment of the so-called Sulphur Commission in 1986. It was a working group, consisting of experts in the fields, that was appointed by the Ministry of Environment and Natural Resources for the development of a new air protection strategy. The Sulphur Commission investigated the possibilities to reduce Poland's emissions of sulphur dioxide by 30 per cent of the 1980 level by the year 2000, seven years later than the requirements of the sulphur protocol. The goal for sulphur reduction was primarily to be achieved by methods such as coal desulphurization, new combustion processes, flue-gas desulphurization, and lime injection. The cost of implementing the programme was set at 500 billion zloties (at 1988 prices) or approximately twenty-five times the expenditures for air pollution control in the year 1986 (Ågren, 1988).

In a statement made in connection with its second general conference in 1987, the PKE claimed that emission of sulphur dioxide had caused a "disaster" in the forests of southwestern Poland and that forests in 77 per cent of the country's territory were threatened by the same pollutant, that ten million people were living in areas where the concentration of air pollution was above the permissible standards, and that the emissions were causing huge economic losses. In order to change the negative trend the PKE demanded the government take action in the following ways:

1. Poland should *immediately* agree upon 30 per cent reduction of its sulphur dioxide emissions according to the international sulphur protocol of 1985. PKE underscored that the economic benefits from such a reduction outweighed the value of the necessary investment expenditures.
2. Replacement of individual boilers by central heating, closures of certain energy-intensive industries, and increased used of renewable energy sources.
3. Elimination of sulphur by way of coal cleaning.
4. Introduction of fluidized combustion technologies.
5. Introduction of flue gas desulphurization equipment in the power sector.
6. Use of the cleanest coal in the most polluted areas and in the protected areas.
7. Initiation of a regional programme with the GDR and Czechoslovakia to reduce transboundary emissions of sulphur dioxide in the so-called Black Triangle.
8. Various measures to improve the situation in areas affected by forest dieback (PKE, 1987: 52-54).

None of these demands, except the first one, differed fundamentally from the air protection policy that was advocated by the Ministry of Environmental Protection and Natural Resources. Hence, the public and governmental agendas with respect to air pollution from stationary sources were almost identical at this time.

The 1988 "National Environmental Programme to the Year 2010", elaborated by the Ministry of Environment and Natural Resources, attached priority to the abatement of sulphur dioxide. The major components in its air protection policy were adoption of coal cleaning technologies, improved combustion processes, and the installation of pollution control equipment such as flue gas desulphurization. The government identified a number of areas where progress was considered to be required:

- elimination of energy-consuming and severely polluting technologies;
- development of pollution control manufacturing;
- expansion of district heating systems;

- modification of transport systems; and
- international cooperation within the framework of the LRTAP and bilateral cooperation with Czechoslovakia and GDR (World Bank, 1989f: 31-32).

In the second part of the 1980s a small group of environmental economists studied the merits of emission trading and the applicability of this policy instrument in a socialist economy. They arrived at the conclusion that the instrument could be applied in Poland (Żylicz, 1998). In 1987 the environmental magazine *Aura* devoted one article to emission trading (Malawski, 1987).

In can be concluded that air pollution from stationary sources was a priority problem for the Environmental Protection Advocacy Coalition in the 1980s. Air pollution was a prominent issue on both the governmental and public agendas. The key actors in the policy process were the Ministry of Environment and Natural Resources and the Polish Ecological Club. The policy process was also inspired by cooperation with foreign governments and environmental organizations.[225]

The 1990s

In 1991 Poland signed a protocol, within the framework of the LRTAP, on emissions of volatile organic compounds (VOC). In 1994 the Second Sulphur Protocol (SSP) within the LRTAP was agreed upon in Oslo[226] that was immediately signed by Poland (Wuester, 1996).

Poland's air protection policy in the 1990s is affected by the result of the ongoing harmonization of Polish environmental law to EU directives. Most importantly, this process implies that technical standards, based on principles such as BAT and BATNEEC, will guide the permit process instead of environmental quality standards.[227]

Throughout the 1990s the main advocates for the introduction of emission trading have been the ecological economists in the Efficiency Advocacy Coalition. In the last few years there is also strong support for this instrument from the energy sector. Organizations that have spoken out in favour of emission trading are Polish Power Grid (*Polskie Ścieci Elektrynycnej* or PSE), the Chamber of Commerce for Energy and Environmental Protection (*Izba Gospodarcza Energetyki i Ochrony Środowiska*), and the independent research institute Energopomiar (Kochel and Pinko, 1996). In addition, a number of representatives of individual power plants are in favour of emission trading schemes on a regional or national level.[228] They are all convinced that emission trading is a cheaper way of meeting their national and international air protection obligations than the present system. Remarkably, in 1993 the power sector, on its own initiative, prepared a draft Air Protection Act containing a provision for marketable permits (Żylicz, 1998). The act was not approved by the parliament.

---

[225] Personal communication, 1995.

[226] A major feature of this protocol is that the environmental targets are set in relation to critical loads (Wuester, 1996).

[227] Personal communication with Jerzy Sommer, 1996.

[228] Personal communications with Sebastian Pejm, Stanisław Gołąb, Jan Rogóż, and Janusz Wójcik, 1997.

In the 1990s industrial branches such as chemistry and steel have frequently interacted with environmental policy-makers to achieve reductions in the fee level. This confirms the conclusion that was drawn in Chapter 5 that interaction between the environmental policy subsystem and the industrial subsystem has increased in the 1990s.

The environmental NGOs have been less active in the debate about air pollution from stationary sources in the 1990s than they were in the 1980s.[229] There are two major explanations for this. Firstly, their focus has switched from stationary to mobile emission sources. For instance, the PKE, the Green Federation and many other organizations have organized manifestations against the plans to establish motorways. Another concern is waste incineration; various environmental organizations have organized campaigns against the construction of waste incineration plants (Karaczun, 1994: 56-57). Secondly, the NGO-community tends to perceive the problem of air pollution from stationary sources to be effectively dealt with by the environmental administration. As one member of the PKE put it:

> I think that there is a general will in Poland, including a genuine will within the Ministry of Environment, to reduce sulphur dioxide emissions. It is the only substance, for which I observe a real will to achieve substantial reductions.[230]

In the environmental NGO community, most attention to the problem of air pollution from stationary sources has been paid by the PKE branch in Katowice. With the support of the International Secretariat on Acid Rain, based in Sweden, this PKE branch established the Information Centre on Air Pollution in 1990.[231]

6.3. Major policy issues

6.3.1. Standards and permits

The permit process is guided by national environmental quality standards (Sommer, 1993: 4). The regulation of the Council of Ministers of 30 September 1980 distinguishes various levels of permissible concentration in the air depending on the area, dividing them into protected and particularly protected areas. The second category comprises health resorts, national parks, nature reserves, landscape parks and other forms of protected areas, and protected forests (Karaczun, 1993: 28).

The 1990 Ordinance on the Protection of Air against Pollution, elaborated by the Ministry of Environment, revised the air quality standards for forty-four air pollutants. It also established permissible deposition rates for cadmium, lead, and dust and established nationwide emission standards for sulphur dioxide, nitrogen oxides, and particulates from fuel combustion (OECD, 1995a). The 1990 Ordinance on the Protection of Air against Pollution is the single most important air protection

---

[229] A case in point is that the Institute for Sustainable Development sent a questionnaire to forty-two environmental organizations to investigate their air protection activities but received not a single response (Karaczun, 1994: 55-56).

[230] Personal communication with Zbigniew Karaczun, 1996.

[231] Personal communication with Aleksandra Chodasiewicz, 1996.

regulation in Poland in the 1990s.

Calculations made by the World Bank showed that it would cost $6-9 billion for fifty-five power stations to comply with the new emission standards of 1998 if end-of-pipe technologies – i.e. flue gas desulpurization equipment –were applied only. With coal enrichment this cost would become considerably lower (Karaczun, 1993: 44). OECD (1995a: 41) estimated the compliance costs at $3-6 billion.

The system of permits in Poland is quite different from the one that is most commonly used in western European countries; Polish permits define an emission limit attuned to certain environmental quality standards while western European permits tend to determine the type of technology to be applied in order to reach a particular limit. Hence, in contrast to their colleagues in western Europe, Polish environmental policy-makers do not put emphasis on *how* the emission limit is reached.

The basis for the Polish air pollution control system is individual permits. The environmental departments in the voivodships issue permits that impose emission standards for a particular source of pollution. According to the 1980 Environmental Act, all plants causing air pollution should have an air pollution permit. It is required that individual air emission limits are stated separately in the permit for each stationary source of air pollution within a plant. On the basis of documentation from the polluters, time limited permits are issued by the regional governors (Jendrośka, 1990: 27).[232]

Polluters who have not been granted permits are obliged to pay fees that are 100 per cent higher than the fees normally paid. Double fees must be paid for the whole air pollution emitted (Sommer, 1993: 57).

Ever since the beginning of the 1980s the voivodship authorities have had serious difficulties in issuing permits to all registered polluters. In 1986 only 34 per cent of the 35,000 registered air-polluting enterprises had any permits (World Bank, 1989f: 41). In 1992 the situation had improved but still only 60 per cent of 46,000 registered air polluters had valid pollution permits (OECD, 1995a: 95).

6.3.2. Emission trading

There are no legal provisions for emission trading in Poland. Despite this, a project with a modified form of emission trading did *de facto* take place in Poland in the early 1990s.

The Chorzów project

As early as 1990 the Economics Department of the Ministry of Environment had begun to search for a site to locate an emission trading experiment. This initiative met with great interest from several US-based organizations, among them the Environment Defense Fund (EDF). The Economics Department and EDF reached an

---

[232] Data submitted by the plants must comprise: (1) technology used in the installation; (2) characteristics of every source of emission; (3) working time of each source of emission during the year; (4) amount and kinds of pollutants in tonnes per year, in kilograms per hour, in grams per second, and in kilograms per production unit for each source; (5) characteristics of the cleaning installations and their effectiveness; (6) conditions of introduction of pollutants to the air; (7) current level of pollution and the future level after the setting into operation level of the plant; (8) maximum concentration of pollutants (in time and space); (9) conditions which influence the spread of pollutants; and (10) plans for air pollution reduction (Sommer, 1993: 52-53).

agreement to develop the project jointly (Żylicz, 1998: 6). In 1991 the EDF and representatives of the Economics Department approached the authorities in Katowice voivodship and Chorzów, a city situated in Upper Silesia, with a proposal to test an emission trading scheme in Chorzów. The city was selected because, firstly, an environmental programme for the most polluting industrial plants had already been proposed and, secondly, the environmental situation in Chorzów had reached dramatic dimensions (Dudek *et al.*, 1992: 9). Three parties of the original programme – the Kościuszko steelworks, the "Chorzów" heat and power plant, and the city authorities – declared that they were interested in participation.

The first step that was taken in the project was that Chorzów city partly covered the costs of the construction of a gas supplying station for the steelworks. This allowed the steelworks to modernize certain technological processes (for instance, the input to the rolling mill) and to close down others (in 1993 the notorious Marten melting stoves were completely closed). In this way the major environmental problems had been eliminated. Subsequently the steelworks turned its attention to its old and polluting heating plant. It decided to close this plant after the Voivodship Environmental Fund and city authorities of Chorzów had agreed to co-finance the construction of a connection between the steelworks and the main thermal pipeline of the heat and power plant. This investment had a positive spin-off effect in that it enabled an expansion of the district heating system, something that meant that a number of local boilers in the city could be taken out of operation.

The trading element involved in the Chorzów project was that the grant (ECU 223,500) that was given by the voivodship environmental fund to the steelworks for its connection with the pipeline actually originated from the power plant. Besides paying normal fees for its air pollution emissions (sulphur dioxide, nitrogen oxides, and particles) to the environmental fund, it added a sum that covered the fund's expenses for the grant to the steelworks. In this way only 40 per cent of the heat and power plants' payment was visible on the account of the environmental fund. The heat and power plant had direct financial incentives to go ahead with the transactions because it had been promised a favour in return: to have its non-compliance fees for excessive dust emission waived (Cracow Academy of Economics, 1995).

The Chorzów project is considered to be a success since it brought about benefits for all parties involved. The voivodship environmental department in Katowice claims that the project speeded up the emission reduction by two and a half years as compared to the normal scenario.[233] Furthermore, the steelworks was modernized (and perhaps avoided a bankruptcy) and the city of Chorzów eliminated several sources of low emission and reduced its costs of hot water transmission from the power plant because the pipeline could use a shortcut through the steelworks. The heat and power plant avoided paying non-compliance fees for its dust emission, it was allowed to postpone expensive dust reduction investment, and it expanded the district heating in Chorzów (Cracow Academy of Economics, 1995: 190-191). The project also increased interest in the use of emission trading. After the project had been launched a number of municipalities in Upper Silesia, for example, Gliwice and Zabrze, approached the voivodship environmental administration in Katowice asking for an approach similar to the one that was developed in Chorzów. However, they were refused because the legal grounds for that kind of treatment were not in place.[234]

[233] Personal communication with Tomasz Żylicz, 1996.

[234] The project in Chorzów was not legal in a formal sense because the environmental framework had not been amended to allow for emission trading. Nevertheless, it was supported by the

The Chorzów project is usually referred to as an experiment. However, the architects of the project rather see it as a demonstration project, an integrated part of an effort to create a constituency for cost-efficient solutions in Polish environmental policy. As Żylicz[235] puts it: "I was absolutely positive that there was nothing to experiment about, just something to demonstrate".

Other preparations for emission trading

Partly as a result of the previous experience in the Chorzów project, a new project was launched by the Ministry of Environment in 1994. The project was delegated to Atmoterm, a Polish firm which formed a consortium with several other consultants and was granted a contract by PHARE to design a tradeable emission permit system for sulphur dioxide and stimulate the implementation of a pilot project in Opole voivodship in southwestern Poland. A precondition for the implementation of the Opole project is that the environmental framework law will admit emission trading (Pazdan, 1996: 14-15).

In 1994 the Ministry of Environment initiated a study devoted to the development of methodologies for assessing costs of meeting selected international environmental standards.[236] The cost for the power sector between 1993 and 2010 for complying with the new domestic emission standards required from 1998 was estimated at $2,397 billion (Cracow Academy of Economics, 1996: 7) while the costs for the same sector to comply with the reduction targets of the SSP by the year 2010 were calculated at $14,647 billion (*ibid.*: 8).

The study concluded that the cheapest way for stationary combustion plants in Poland to meet the requirements of the SSP is to initiate a programme for banking and trading in sulphur dioxide emission permits (*ibid.*: 21). The intended effect of such a scheme would not only reduce the costs but also increase the likelihood of reaching compliance with Poland's commitments within the SSP. The rationale for using an emission trading scheme for the power sector's compliance with the SSP was formulated in the following way:

If banking and trading in emission permits is allowed up to the year 2010, each rationally operating enterprise will choose its own deadline for implementing the programme to bear the least costs of the project to reduce sulphur dioxide. If, however, trading is prohibited, we assume that each plant will have to achieve additional levels of emission reduction at regular intervals, which would be set out by the government in a schedule for reduction of emissions and translated by provincial governors into individual decisions on emission limits. This means that some plants would have to make capital outlays for pollution control equipment earlier than would be optimal (least-cost) for them. In the narrow sense of the environmental effectiveness it could be beneficial. For most plants, however, it would entail additional costs (*ibid.*: 10).

Minister of Environment, Maciej Nowicki, who wrote a letter to the Katowice environmental administration and authorized them to proceed with the project (personal communication with Tomasz Żylicz, 1996).

[235] Personal communication, 1996.

[236] The study is entitled Development of Cost Methodologies and Evaluation of Cost-Effective Strategies for Achieving Harmonization with EC Environmental Standards.

According to the proposed scheme, the environmental administration should give the plants in the programme certificates of emission reduction credits (i.e. unused permits), when they can prove that they have complied with more stringent emission conditions than the standards binding it in a given period. Plants, which have legally certified decision on more stringent conditions for emission, may either retain unused credits in order to use them at a later stage (but not later than 2010) or sell unused credits to other plants (*ibid.*: 22):

Banking certificates and their tradeability would provide plants with an incentive to implement the emission standards of the Second Sulphur Protocol earlier, because the reductions obtained ahead of the schedule in the Protocol would create prospects of additional monetary income from sales of unused certificates. For plants, in which meeting the standards in short time would entail very high costs, the suggested programme would make it possible to legally purchase rights to temporarily exceed the standards and to postpone reduction of emissions till the later years of the programme. Very old plants would be able to buy rights to the time of their closure (*ibid.*: 23).

The target group for the programme, proposed to last between 1998 and 2010, is large and medium-sized stationary combustion installations (*ibid.*: 22).[237]

The SSP recommends that the countries that have signed the convention adopt new emission standards in the year 2004. The study discovered that moving the introduction of those standards to the year 2010 would entail cost savings equivalent to more than $4 billion for the large power plants in Poland (*ibid.*: 9).

It was proposed that, for stacks that are lower than 100 metres, the permitting authorities at the voivodship level should be given the right of veto against purchasing of permits if the stacks constitute a threat to the local ambient air quality.[238]

The future of emission trading in Poland lies in the hands of the Ministry of Environment. In 1991 it had a Minister, Maciej Nowicki, who belonged to the EAC and who was personally convinced that the tool should be introduced as soon as possible. The positions of the following Ministers – Kozłowski (the GAC), Hortmanowicz (the MAC), Żelichowski (the MAC), and Szyszko (the MAC) – have been more hesitant. According to Radosław Gawlik[239] of the parliamentary committee for environmental protection, the Ministry of Environment has never given a clear signal to the parliament about what role emission trading should play in the future environmental management system in Poland: "there is a kind of chaos; some say this, others say that. It is not clear to me what policy is advocated by the Ministry of Environment".

In June 1996 the Ministry's position was explained by deputy minister Krzysztof Szamałek (MAC) at an international workshop on marketable permits in Jadwisin near Warszawa. He made clear that the Ministry is determined to introduce emission trading: "The question is not whether but rather how and where to apply this

[237] The policy-makers are open to two options: one with plants of 50 MW or more, one with plants over 0.2 or 1.0 MW. The former is likely to lower the number of transactions and the differences between the marginal costs of emission reduction among power plants. The latter may increase the administrative costs of operating the emission permits market (*ibid.*: 22).

[238] Personal communication with Grzegorz Peszko, 1996.

[239] Personal communication, 1997. After the parliamentary elections in the autumn of 1997 Gawlik became the first deputy Minister of Environment.

instrument. It is only a matter of pace and scope" (Szamałek, 1996: 12).

6.3.3. Enforcement

The air protection policy is enforced by PIOŚ. The main legal instrument that belonged to the Inspectorate in the 1980s was an order issued after a control of the industry in question. This order was binding on the controlled plant but the Inspectorate had no right to execute it because it had no right to stop environmentally harmful activities but had to ask the voivodship authorities for it (Jendrośka and Radecki, 1991). Until 1991 fines and closures were instruments in the hands of the voivodship environmental departments. This was changed in 1991 by the Statute on the State Inspectorate for Environmental Protection, which transferred these enforcement instruments to the inspectorate.[240]

The most common enforcement measure in the Polish air protection policy is a fine for non-compliance with the requirements defined in the pollution permit. The genesis of this instrument goes back to the 1960s. The present fine system is regulated under the 1980 Environmental Act and in a 1987 Executive Order (amended in 1991 and 1992) (Jendrośka, 1990: 27). In the first half of the 1980s fines were four times higher than the normal fees (World Bank, 1989f: 41). Since 1987 the rates of the fines are ten times higher than the rates of the fees.

The second major enforcement instrument in the hands of the Inspectorate is closure. PIOŚ has two legal possibilities to close down facilities. The first possibility is to apply Article 31 in the environmental protection framework act, which states that if a facility emits more than its permit allows for, and the plant fails to abate the excessive pollution, then PIOŚ has the right to close the plant. The second possibility is offered by Article 81 in the same act, which states that if a clear-cut negative impact on the environment or public health due to the activities of a particular facility can be proved, then PIOŚ *must* close it. An important difference between Articles 31 and 81 is that in the former a precondition for closure is that the polluting plant in question has a valid permit, while in the latter this is not required.[241] Article 81 is rarely applied. A major reason for this is that in industrial areas with many emission sources it is difficult to identify the emission source that is causing the damage.[242]

In 1996 PIOŚ' enforcement power was somewhat weakened because the government decided that PIOŚ cannot decide about a closure without having consulted the voivodship authorities.[243]

6.3.4. Fees and financing

The 1980 Statute on the Protection and Shaping of the Environment introduced fees for emission of air pollution. The fee system of the 1980s was accompanied by two problems. First, the fee level was too low to constitute an incentive for the polluters to change their behaviour. For example, of 1,400 enterprises surveyed in 1987, air

[240] Personal communication with Wojciech Radecki, 1996.

[241] Personal communication with Wojciech Radecki, 1996.

[242] Personal communication with Wojciech Radecki, 1996.

[243] Personal communication with Wojciech Radecki, 1996.

pollution fees and fines (together with other charges) amounted to 0.2 per cent and 0.3 per cent, respectively, of total costs. The low rates of the fees were partly related to the fact that there was no inflation indexation of the fees in the 1980s (World Bank, 1988f: 41).

In 1995 charges were levied on sixty-two atmospheric substances. The charge levels were strongly differentiated. The highest rates apply to the most toxic emissions such as benzene and chromium. The charges for sulphur dioxide and nitrogen oxides are among the highest in the world (Śleszynski, 1996a). For example, the charge for sulphur dioxide ranked third in the world after the Swedish and Norwegian ones (Stavins and Żylicz, 1995: 4). The air pollution fees are earmarked for the environmental funds with the following distribution: 36 per cent to the National Fund for Environmental Protection and Water Management, 54 per cent to the Voivodship Funds for Environmental Protection and Water Managament, and 10 per cent to the Municipal Environmental Funds. The fees for nitrogen oxides constitute an exception from this rule since 90 per cent of this fee is earmarked for the National Fund while the remaining 10 per cent goes to the Municipal Environmental Funds (Anderson and Fiedor, 1997: 5).

In general, it is a mixture of soft loans from the National Fund for Environmental Protection and Water Management and commercial credits from banks that finances the largest air protection investments. Most of the power plants that have invested in air protection equipment in the 1990s have signed long-term contracts with the Polish Power Grid which guarantees that they will have a stable income over at least one decade which will be enough to pay back the loans.[244]

6.4. Implementation

After the demise of socialism in 1989, Poland's air pollution emissions have fallen sharply. Total sulphur dioxide emissions fell from 4,180 million tonnes in 1988 to 2,645 million tonnes in 1994 or 37 per cent. The first reduction goal for this substance established by the Polish government in the national environmental policy of 1991 (30 per cent of the 1980 level by the year 2000) was reached in 1992. In 1994 the emissions were 34 per cent below the emission level in 1980 (Salay, 1996: 25).[245]

Table 6.3 shows that the annual sulphur dioxide emissions have declined more rapidly than the decline in GDP and industrial production. The decline in economic activity may only provide part of the explanation. Salay (1996) argues that the switch to coal with lower sulphur and ash content is an important factor for explaining the fall in air pollution. In his view, there were two driving forces behind this switch. The first one was the restructuring of the power industry,[246] the introduction of hard

---

[244] Personal communication with Roman Janiczek, 1997.

[245] The concentration of sulphur dioxide in Kraków was 9.5 per cent higher in 1991 than in 1990. The ambient air quality was twice as high the permissible standards. At the same time, industrial emissions had decreased. The deterioration in air quality was related to increased use of low-quality coal, which in turn was related to a liberalization of the trade of hard coal. Before 1990, Kraków was only allowed to burn high-quality coal in the local boilers (Karaczun, 1993: 34).

[246] The former centralized authority in charge of electricity generation, transmission, and distribution was split up into separate companies with the aim of creating financially independent joint stock companies owned by the state. An immediate consequence of this change was that the power

budget constraints, and coal price increases. The second driving force was, according to Salay, environmental policy reform that, *inter alia*, implied higher emission fees and the imposition of new standards. Hence, it was the combined effect of restructuring and environmental policy reform which created incentives for power plants to switch to high-quality coal (*ibid.*: 11).

According to the power plants themselves, there have been several driving forces for their ambitious air protection investments.[247] First, the new emission standards that the Ministry of Environment introduced for combustion processes in 1990. A second driving force has been Poland's international obligations. The Polish Power Grid has been connected with the western European power grid UCPTE that allows Poland to export electricity. One of the requirements for this export is that the energy sold is produced in a way that fulfils EU environmental requirements, such as the 1988 Directive on Large Combustion Plants, which obliges large power plants to reduce certain acidifying substances (Liefferink, 1995: 115). Poland's commitment to comply with the reduction targets of the Second Sulphur Protocol has also been important.

Table 6.1 Annual percentage changes in GDP, industrial production and air pollution emissions in Poland, 1990-93

Year	GDP	Industrial production	Dust	Carbon oxides	Nitrogen oxides	Sulphur oxides
1990	-12	-24	-23	-17	-18	-21
1991	-7	-12	-21	-28	-6	-8
1992	2	4	-26	-15	0	-10
1993	4	7	-13	-21	0	-3

Source: Cole, 1995b: 283.

It should be mentioned that many power plants were put on the list of the eighty largest polluters. They are determined to leave this list as soon as possible to improve their environmental image, both in Poland and abroad.

The main technological pillars in the Polish air protection policy have been coal cleaning, installation of flue gas desulphurization equipment at large power plants, introduction of fluidal boilers in the energy and industry sectors, and conversion of coal fired boilers to gas. In the 1990s virtually all large coal fired power plants in Poland have launched programmes for investment in advance equipment to further reduce their air pollution emissions. Flue gas desulphurization equipment (FGD) is already in place at Bełchatów and Turek, Poland's largest power plants fired by brown coal (Cole, 1998), and at Jaworzno III, a hard-coal fired power plant. Power plants such as Konin, Łagisza, Łaziska, Rybnik, Siersza, Pątnów, and Połaniec are all in the phase of installing FGD equipment that will bring down emissions significantly.[248]

plants were empowered to make their own decisions on issues like buying fuel and selling its power to the grid (Salay, 1996).

[247] Personal communications with Stanislaw Gołąb, 1997; Sebastian Pejm, 1997; and Janusz Wójcik 1997.

[248] Personal communications with Wojciech Jaworski, 1995; Anna Zawiejska, 1996; Stanislaw Gołąb, 1997; and Sebastian Pejm, 1997.

6.5. Advocacy coalitions and their belief systems

The conclusions drawn about advocacy coalitions in Chapters 4 and 5 on national environmental policy are also valid with respect to the issue of air pollution from stationary sources. Hence, the issue was dealt with by the Environmental Protection Advocacy Coalition from the mid-1980s to the early 1990s. After that, three advocacy coalitions appear: the Mainstream Advocacy Coalition, the Green Advocacy Coalition, and the Efficiency Advocacy Coalition.

Below follows a reconstruction of the coalitions' beliefs systems with respect to air pollution from stationary sources. Only the secondary aspects are addressed below because the beliefs on air protection do not relate to the entire environmental policy subsystem.

The Mainstream Advocacy Coalition (secondary aspects beliefs)

Permits. The MAC considers the permit system to be cumbersome to administer. For instance, MAC representatives of the environmental protection departments at the voivodship level complain that they are overloaded with work. As a consequence, there is not sufficient time to register new firms that are established and there is not even time to deal with the existing polluters.[249]

The MAC[250] believes that the permit system forces the environmental departments to set unrealistic deadlines for the permits.

There is, according to the MAC, a lack of consistency between various types of permits (for instance, planning permissions, building permissions, water permissions, air pollution permissions, noise and vibration permissions) that are issued. This is a major deficiency of the Polish legal system for environmental protection (Jendrośka and Radecki, 1991).

The MAC is not against emission trading but the coalition is still hesitant about this approach for three major reasons. Firstly, the MAC considers the financing system in Poland to function very well and is therefore unwilling to accept a major reconstruction of this system because of an introduction of emission trading. Secondly, the MAC wants to know precisely how marketable permits relate to the EU's new directive on Integrated Pollution Prevention and Control that is supposed to come into force after 2006. Thirdly, the MAC is not sure whether the voivodship environmental departments and PIOŚ are sufficiently prepared to effectively monitor permit transfers (Szamałek, 1996).

Standards. Representatives of the MAC claim that the way ambient quality standards are translated into emission permits is unfair to industry. First of all, only a few cities (such as Kraków, Katowice, and Szczecin) have continuous monitoring of the air quality. In other places the pollution concentration is measured with equipment which often is not standardized, something which leads to large differences in results. Another complaint is that the dispersion models used by the voivodship authorities are not accurate. This is particularly evident in the case of simulations about air pollution dispersion from high stacks. The models fail to account for real meteorological conditions and for the role of background concentration. Moreover, a particular emission source's impact on the local air quality is only roughly estimated. Industries

[249] Personal communications with Anna Zawiejska and Jan Dudzik, 1996.

[250] Personal communication with Wojciech Jaworski, 1995.

with high stacks are treated in an unfavourable way because they tend to be made accountable for both local and distant concentrations. The consequence of this discretion exerted by the environmental departments is that industry is punished in the form of strict emission permits (Cracow Academy of Economics, 1995: 114).

Enforcement. The coalition argues that the fine is unquestionably a proper instrument for some polluters but for others, predominantly companies in a poor financial situation, it does not change anything. These polluters tend to remain in non-compliance. In order to bring these firms into compliance a new and more flexible enforcement instrument of environmental law, *viz.* Negotiated Compliance Schedules (NSC) is seen as necessary.[251] According to PIOŚ, "compliance programmes are to be a legal form of introduction of legal instruments, aimed at ensuring gradual achievement of generally binding standards by enterprises (legal entities) causing an adverse impact on the environment and not complying with the requirements currently in place (...)". Furthermore: "An enterprise which is not financially able to meet the requirement immediately, will now have the chance to do it gradually in a precisely defined time framework and using solutions, formulation of which it can influence to some extent" (State Inspectorate, 1996: 8-9).

Technology. End-of-pipe technologies should be an important element of the air protection policy, according to the MAC.

The Green Advocacy Coalition (secondary aspects beliefs)

Permits. The Green Advocacy Coalition has not been involved in the ongoing debate on emission trading and its knowledge of the policy instrument is rather limited. Those who have raised objections against the introduction of the tool have generally claimed that emission trading is a too explicit price for pollution. They have also argued that the idea of transferring permits is not consistent with a basic principle in Polish environmental law, *viz.* that no administrative decisions should lead to a deterioration of environmental quality. Nevertheless, there are a few environmental NGOs that are in favour of the tool, for example, the Institute for Sustainable Development and the Foundation for Effective Energy Use (*Fundacja Efektywnego Wykorzystania Energii* or FEWE).[252]

Enforcement. The main standpoint of the GAC with respect to enforcement is that the largest polluters are effectively taken care of by the PIOŚ and the voivodships' environmental departments. This trust in the present enforcement mechanisms may explain its silence in the ongoing debate about enforcement of environmental law. However, on one occasion the NGO community made a common statement related to enforcement. In a NGO report on sustainable development, prepared jointly by the Polish environmental organizations in connection with the UNCED in Rio de Janeiro in 1992, the link between effective enforcement and increased access to information was emphasized (Institute for Sustainable Development, 1992: 7-8):

One of the major goals of the state's ecological strategy should be to allow the public a much

---

[251] Personal communication with Zbigniew Kamieński, 1996.

[252] One interviewee representing the NGO community argues that environmental organizations should be able to buy permits in order to refrain from using it (personal communication with Robert Alberski, 1996).

greater access to information on the condition of the environment. The government should submit regular reports and continue collecting and publishing data on industrial plants that are the largest polluters, particularly those whose levels of emissions significantly surpass admissible standards. Lists of such industrial polluters should be publicized regionally and nation-wide and should serve as a starting point for individual projects undertaken with a view to improving the condition of natural environment.

Radosław Gawlik[253] of the GAC argues that enforcement would benefit from stronger environmental organizations because the pressure they exert increases the effectiveness of the environmental administration. Gawlik is in favour of Negotiated Compliance Schedules on the condition that the negotiations will be controlled by the public: "If environmental organizations will be able to supervise the process I am not worried."

Fees and financing. According to the GAC, the National Fund devotes too little of its resources to air protection. It points to the fact that more than half of the Fund's income originates from air pollution fees but only about 30 per cent of the disbursements were assigned to air protection investments (Karaczun, 1994: 54). The GAC is convinced that environmental effectiveness should be more important for the environmental funds than cost-effectiveness (*ibid.*: 46).

Technology. End-of-pipe technologies are firmly rejected by the GAC. It argues that Poland should shift its focus to clean and energy efficient technologies and renewable energy sources which are perceived to promote sustainable development in a better way than end-of-pipe equipment. This was also underscored in the joint NGO statement before the UNCED:

Priority should be given to those technologies, which rely on renewable natural resources and do not depend on external markets. One of the major tasks is to eliminate energy and material-consuming technologies. It is necessary to promote those technologies, which decrease the consumption of energy and raw materials, rely on recycled materials, close circuit production, whose waste is biodegradable, and suitable for further processing and which do not bring about the growth in annual use of power and raw materials (Institute for Sustainable Development, 1992: 19).

Furthermore:

(...) to check energy wastefulness is a major and indisputable challenge that all Poles are facing. Saving on energy will be the same as increasing energy supply, because more economical energy consumption increases a stream of useful energy that people need. More economical and more efficient use of energy will be possible on the condition that long-term and comprehensive programmes are elaborated. The same holds for the search for new sources of energy or greater exploitation of the existing ones. It is especially in Poland's case that energy savings will be the safest, the cheapest, and the most profuse source of energy for many years to come. The energy saving strategies will call for a great involvement on the part of society (*ibid.*: Appendix 4: 1).

The Efficiency Advocacy Coalition (secondary aspects beliefs)

Permits. The central EAC argument against the present permit system is that the environmental departments are given very little room to diversify requirements

---

[253] Personal communication, 1997.

according to the abatement costs. In other words, the present system does not promote cost-effective allocation of abatement costs across emission sources:

Once issued, permits cannot be taken away from the sources where abatement is cheap (so emission can be easily reduced) and transferred to high cost polluters even if it had no impact on ambient quality. As permits are specified not only for the plant as a whole but also for each stack it is impossible – without administrative decision – to reallocate emission from one stack to another even within the same plant (Cracow Academy of Economics, 1995: 113).

Emission trading offers a possibility to overcome this problem and should therefore be widely used.

The EAC claims that the permit system in its present form is cumbersome to administer. Wojciech Beblo[254] of the EAC argues that as a director of the voivodship environmental department in Katowice between 1991 and 1995 he had to deal with six different regional agencies dealing with environmental protection, which made it difficult to coordinate the air protection policy.

Another argument against the present permit system, raised by representatives of the EAC (*ibid.*:113), is that it forces the environmental departments to set unrealistic deadlines for the permits. They are not satisfied with the fact that the permit must be obeyed immediately after the administrative decision becomes effective, not after it is possible to install abatement equipment.

The EAC points out that the present permit system makes it difficult for the environmental departments at the provincial level to reconcile environmental goals and developmental goals. Usually, the environmental departments have distributed all permits to the existing polluters in the region. Thus, when newly established firms ask for permits, the departments sometimes refuse to issue new permits, by referring to the fact that each additional permit issued would lead to a violation of the national ambient quality standards. In other instances, when the voivodship authoritities do not want to stop the establishment of new plants, they simply issue permits and consciously violate the air quality standards.[255]

Standards. The EAC argues that the ambient standards introduced in 1990 are too strict. It points to the fact that Poland's ambient standard for sulphur dioxide (32 $\mu g/m^3$ mean annual concentration) exceeds the standards in the European Union (where 80-120 $\mu g/m^3$ apply), Switzerland, and the USA (Żylicz, 1995a). A lack of compliance with the present Polish standards may undermine the authority of environmental administrators.

An important barrier for an effective air protection policy is the lack of responsibility for meeting ambient standards in that the environmental administration at voivodship level is not made accountable for the attainment of standards. Żylicz contrasts the Polish approach with the US EPA policy which early in the 1970s made each governor responsible for meeting the federal ambient standards. As a result of this responsibility, he argues, they discovered that emission trading is the most proper instrument to improve ambient air quality.

Enforcement. The EAC is strongly opposed to Negotiated Compliance Schedules, which is advocated by the MAC (see below) and advances three major arguments against that approach. Firstly, there will be too much room for discretion in the

---

[254] Personal communication, 1995.

[255] Personal communication with Grzegorz Peszko, 1996.

negotiations. Secondly, there is an obvious risk for bribes and informal rewards when underpaid representatives of the environmental administration are negotiating with rich and large enterprises. Thirdly, the institutional capacity of the environmental administration is not sufficient for the task: "The system of individual negotiations may work if the public administration is accountable, transparent, and extremely strong. The Polish public administration is neither accountable, nor transparent, nor strong".[256]

Fees and financing. The EAC points out that the fact that industry pays most of the charges while the municipal sector receives most of the revenues is doubtful in several respects. Firstly, it is not clear that the maximum environmental benefits are to be achieved in the municipal sector. Secondly, giving the municipal sector a lot of money can work as a strong disincentive to operate the environmental infrastructure in an efficient way. Thirdly, there is a risk that industry may start to argue that it is paying too much but receives to little and to begin to question the rationale of the financing system.[257]

The EAC also argues that it is not clear whether the fees are supposed to play an incentive or revenue-raising role, which leads to a situation where the fees do not play either of these roles very well. The fee system is, in the EAC's view, overly complex, and this results in increased administrative costs and a low degree of implementability (Cracow Academy of Economics, 1995: 115).

The main beliefs identified in this chapter are summarized in Table 6.2.

Policy-oriented learning

A considerable amount of learning can be observed in Poland's air protection policy. It is noteworthy that most learning took place *within* the EPAC. A major learning process *across* coalitions that thus far could be observed in the 1990s is that the MAC seems to be open for a limited use of marketable permits, a policy instrument advocated by the EAC, *on top* of the existing policy instruments that are in use in Poland. There are also a few organizations within the GAC that accept marketable permits.

Learning has been facilitated by the fact that air pollution has been on top on the environmental agenda since the mid-1980s and that two of the Ministers of Environment (Stefan Jarzębski and Maciej Nowicki) were air protection experts. Furthermore, international cooperation has promoted learning. Poland ratified the LRTAP in 1985 and after that Poland has signed and ratified a number of protocols under this convention. In the 1980s there were more and more contacts and cooperation between Polish and foreign experts (biologists, ecologists, engineers etc.) on air pollution in general and acid rain in particular. The Polish experts had become active participants in the "epistemic community" dealing with air pollution. Epistemic communities are, in Haas' (1990: 349) definition, "transnational networks that are both politically empowered through their claims to exercise authoritative knowledge and motivated by shared causal and principled beliefs". Tomasz Żylicz learned about emission trading as a guest researcher in the USA in 1977-78 and 1988-89, and brought this idea to Poland. In the latter part of the 1980s contacts between Polish and western environmentalists increased substantially. Organizations such as the

[256] Personal communication with Grzegorz Peszko, 1996.

[257] Personal communication with Grzegorz Peszko, 1995.

International Secretariat for Acid Rain, Friends of the Earth International, *Bund für Naturschütz* from West Germany, and the Swedish-Polish Association for Environmental Protection exposed Polish environmentalists to the acid rain issue (SPM, 1989).

Table 6.2 Belief systems identified in the air protection policy

The Mainstream Advocacy Coalition	The Green Advocacy Coalition	The Efficiency Advocacy Coalition
Secondary aspects		
The current permit system is cumbersome to administer.	It is principally wrong to trade with pollution.	The current permit system is cumbersome to administer.
There are unrealistic deadlines in the permit process.	More public participation and easier access to information will make the air protection policy more effective.	There are unrealistic deadlines in the permits process.
Separate permits should be replaced by integrated permits.	The National Fund spends too little money on air protection.	The present air protection policy creates unnecessary conflicts between air quality and economic development.
The present system for financing air protection investments should not be changed.	Clean and energy efficient technologies and renewable energy sources should be the main pillars of the air protection strategy.	Emission trading should be widely applied because it combines economic efficiency and environmental effectiveness.
The environmental administration is not prepared to effectively monitor emission trading.	End-of-pipe solutions should be avoided or not applied at all.	The present air protection standards are too strict.
The way ambient quality standards are translated into emission permits is unfair to industry.		The voivodship environmental administration should be made accountable for the attainment of air quality standards.
Enforcement would improve by the introduction of Negotiated Compliance Schedules.		The institutional capacity of the environmental administration should be improved.
The application of end-of-pipe technologies is important.		Negotiated Compliance Schedules should not be used because of the risk of bribes, discretion, and the low institutional capacity of the environmental administration.
		It is wrong that industry pays most of the charges and that the municipal sector receives most of the revenues.
		The role of the fees should be precisely defined.

6.6. Summary and conclusions about policy change

Poland's first initiatives to address air pollution from stationary sources were taken by the government in the 1950s and 1960s and focused on dust emissions having impact on local air quality. The first public demands to address the air pollution problem were articulated by the PKE from 1980 and onwards. Air pollution became a priority problem for the government in the second part of the 1980s as the result of the accumulation of scientific evidence about the impact of air pollution and increased public and international pressure to reduce air pollution emissions. In the 1990s emissions of air pollution have fallen sharply.

Ever since 1980 the Polish air protection policy has had the following main pillars:

- individual permits issued by the voivodship environmental departments to the polluters;
- national ambient air quality standards;
- fees for legal emissions and fines (four times the fees; ten times after 1987) for emissions above the permissible levels; and
- environmental funds that use the fees and fines for investments in pollution control.

All of the three advocacy coalitions that were identified in Chapter 5 also appeared in this case study. However, the GAC's main field of interest with respect to air pollution is emissions from mobile sources. Therefore, the discourse is dominated by the EAC and the MAC.

A modified form of emission trading took place in Chorzów in the early 1990s. The main supporters of the instrument are the ecological economists and the power sector, that is, the EAC. It is still an open question whether marketable permits will ever be used on a large scale in Poland. The EAC's failed attempts to introduce emission trading on a large scale can be seen to be a result of the MAC's domination of the key positions in the Ministry of Environment since mid-1992.

Just as in the case of national environmental policy, the policy process had state corporatist features in the 1970s and societal corporatist features in the 1980s. In the 1990s the policy process has the properties of a pluralist one.

Poland's air protection policy has been strongly influenced by international conventions and agreements, both before and after systemic change. Policy-oriented learning within the EPAC in the 1980s and within the MAC and the EAC in the 1990s has been important but has so far not led to any revisions of the major assumptions of the air protection policy.

7. SALINE WASTE WATER FROM THE HARD COAL SECTOR

The topic of this case study is a major implementation failure in Poland's environmental policy in the 1990s: the discharges of saline water to the Wisła and Odra rivers. This book has so far concluded that, since systemic change in 1989, Poland has improved its environmental performance substantially. The mechanisms for (1) the financing of environmental protection measures (notably the National Fund for Environmental Protection and Water Management) and (2) the enforcement of environmental law (executed by the State Inspectorate for Environmental Protection) have eliminated much of the notorious implementation deficit of the 1980s.[258] However, with respect to the discharges of saline waste water from the hard coal mining sector in Upper Silesia in southwestern Poland, these two mechanisms have not worked properly and hence very little progress can be discerned.

Sector 7.1 is devoted to the hard coal sector. Sector 7.2 discusses the environmental problems of the hard coal sector, with a focus on the saline waste water discharges. Section 7.3 analyses the policy processes before and after 1989. Section 7.4 draws attention to four central policy issues, *viz.* standards and permits, enforcement, fees and financing, and technology. Section 7.5 addresses the issue of implementation. Section 7.6 examines the belief system of the actors involved in the salination problem. This section is interested not only in beliefs held by policy actors within the environmental policy subsystem but also beliefs held by the actors in the mining subsystem. In contrast to Chapter 6, this chapter analyses beliefs on both the policy core level and the secondary aspects level. Section 7.6 also deals with policy-oriented learning. Section 7.7 draws conclusions about policy change and summarizes the chapter.

7.1. Introduction

The hard coal sector

The coal sector was a notorious loss-maker in the People's Poland, but due to its strategic importance for the economy it received massive support from the state.[259] A major reason for the losses was that the central planners deliberately kept the wholesale coal price at a low level. They tried to compensate the insufficient revenues by keeping the production costs, for example capital costs, artificially low. Another factor that had a detrimental impact on the economic performance of the sector was the comparatively high labour costs.[260] However, the main barrier for avoiding losses

---

[258] See, for example, OECD (1995b).

[259] As Radetzki (1995: 9) aptly notes, "until 1989, the Polish coal sector suffered from perennial losses, which were covered by equally perennial subsidies from the public budget".

[260] "The cost of labour in a wide sense (...) comprised a plethora of social facilities like housing, schools, hospitals and public transport, provided by the mining companies for free, or at very

was a phenomenon that appeared in all sectors of the socialist economy: the lack of mechanisms for rational allocation of the resources. As Radetzki (1995: 10) clarifies:

Industrial managers during the communist era, including those responsible for hard coal mining, were not required to maximize economic surplus. Instead, their objective was to maximize output, given the amount of resources that they succeeded in extricating from the central planners. Under the circumstances, the managers did not allocate resources in an economically rational manner. But even if they had wanted to, these managers would have been unable to pursue economic rationality, because the prices and costs that they confronted, had been set by the planners, and were, by and large, void of economic meaning.

Since systemic change 1989 output of hard coal has declined sharply (Table 7.1).

Table 7.1 Polish hard coal production
(in million tonnes saleable)

Year	Production	Year	Production
1988	192.2	1991	140.0
1989	177.4	1992	131.3
1990	147.5	1993	130.2

Source: ING BH Consultants *et al.*, 1994: A-4.

This decline has been related to several factors. The most important ones have been the economic slump during the first years of the transition, increased prices for coal,[261] and a loss of the export market to former CMEA countries (Radetzki, 1995: 11). Other driving forces behind the reduction in coal production have been industrial restructuring, competition from other fuels, energy conservation, and environmental policy (ING BH Consultants *et al.*, 1994: 2-2).

Table 7.2 Subsidies as part of coal sales

Year	Percentage
1990	54.6
1991	12.5
1992	1.1

Source: ING BH Consultants *et al.*, 1994: A-19.

A first step towards a reorganization of the hard coal sector was taken in 1990, when the centralized management system was abolished, and the seventy or so mines were transformed into independent enterprises, albeit still owned by the state. The same year the government began to free prices and started to reduce subsidies for mines that did not cover their costs (World Bank, 1993c: 6). Coal subsidies declined from $2.294

low charges. Overemployment was common, with ensuing low labour productivity levels" (*ibid.*).

[261] The coal price on the domestic market has reached world price level (ING BH Consultants *et al.*, 1994: 2-2).

billion in 1989 to $134 million in 1993 (Radetzki, 1995).[262] For comparison it can be noted that the annual German hard coal subsidies amounted to DM 12 billion in the mid-1990s (Böhringer, 1996). In 1992 only 1.1 per cent of the coal sales were subsidized (Table 7.2). The State Agency for Hard Coal (*Polska Agencja Węgla Kamiennego* or PAWK) was created to oversee the mines on behalf of the government. Another important reorganization occurred in 1993 when most of the hard coal mines were grouped in seven joint stock companies. The rationale for this was the need to strengthen the sector administratively and financially, to facilitate further reconstruction, and to avoid the appearance of monopolitic tendencies on the domestic market (Radetzki, 1995: 11). In 1998 PAWK was transformed into the State Agency for Reconstruction of the Hard Coal Sector (*Państwowa Agencja Restrukturyzacji Górnictwa Węgla Kamiennego SA* or PARGWK).

For each year in the 1990s the hard coal sector has become more and more indebted. The financial situation reached such a critical level that the coal sector had to take short-term credits in order to be able to pay salaries and interest on loans taken from banks (Ministerstwo Przemysłu i Handlu, 1996: 10-13). There are a number of factors that had a negative impact on the economic performance of the sector in the 1990s. Firstly, the contracting coal demand in Poland resulted in unfavourable selling prices of coal. The coal mines tried to compensate the losses in the domestic market by increasing exports. However, the export price was lower than the selling price on the domestic market and far beyond the costs of production. Secondly, wages increased more than expected.[263] Thirdly, the coal price increased slower than inflation. For example, selling prices in 1995 increased by 14.9 per cent, while the average increase for industrial products reached 25.4 per cent. The main reason why coal prices have been kept artificially low is that constant increases in coal prices are considered to fuel inflation. As the Ministry of Industry and Trade put it: "[This policy] is justified in macroeconomic terms but it brings the hard coal sector to financial ruin" (*ibid.*: 13). Fourthly, in relation to the złoty the dollar course became lower than had been assumed. This had negative impact on the export price and on the trade relations with the power industry with whom the agreements regulating coal prices were based on the dollar course. Lastly, energy prices have been raised much more than coal prices. This has favoured the power sector on the cost of the hard coal sector.[264]

A governmental policy paper on the reconstruction of the hard coal sector was submitted in October 1995. It was approved by the government in April 1996 (*ibid.*). This programme, colloquially named the Markowski-plan after Jerzy Markowski, the deputy Minister of Industry and Trade in charge of the mining sector between 1993 and 1997, takes into account financial viability, privatization, closures, social issues, infrastructure, and environmental protection.[265] After the parliamentary elections in

---

[262] Subsidies for the following tasks were continued: liquidation of coal mines, social security for the unemployed, elimination of mining damage, and rescheduling of various debts (personal communication with Jerzy Swadowski, 1996).

[263] The salaries increased by more than 14 per cent compared with what had been calculated in the budget (*ibid.*: 12).

[264] Personal communication with Andrzej Szeja, 1996.

[265] On 30 April 1996 the Council of Ministers approved a reconstruction programme for the hard coal sector for the period 1996-2000. According to the programme, production will decrease from 130 million tonnes to some 120 million tonnes in the year 2000, of which 98 million tonnes for sale on

the autumn Markowski's position was taken over by Janusz Steinhoff of AWS. Under Steinhoff's direction the government initiated work on a new reconstruction programme, which is intended to have the status of a law (*Rzeczpospolita*, 1998b).

7.2. Environmental issues

The mining activities in the Upper Silesian coalfield in southwestern Poland, an area representing the greatest concentration of deep mining in the world, have a detrimental impact on the environment in four major ways (World Bank, 1993c: 30).

- Disposal of waste rock. Each tonne of produced hard coal is accompanied by the production of almost half a tonne of waste rock and material.[266] Large amounts of waste also arise in connection with coal preparation. In 1989 most of the waste was disposed of in land levelling and reclamation (43 per cent) and in surface dumps (38 per cent). Some 12 per cent were used for stowing underground. The hard coal sector has problems in finding additional dumping space (World Bank, 1993c: 30-31).
- Surface subsidence. The surface in Upper Silesia is seriously damaged because of various forms of underground movements in the coal mines.[267] The effects of these movements are manifested in the form of cracks in the buildings and damaged infrastructure such as roads, pipelines, electricity cables, and railway tracks.[268] In the mid-1980s 34,000 flats had been damaged (Jurkiewicz and Matiakowska, 1984). Between 1980 and 1986 more than 700 shakes were registered in Upper Silesia (Ciesielski, 1988).

---

the domestic market. At least eleven mines will be completely closed. The decrease in production caused by these closures will partly be compensated by increased production at the most effective mines (Ministerstwo i Przemysłu, 1996: 15-22). Employment in the hard coal sector is supposed to be reduced by 80,000 persons. This measure is considered to be necessary in order to reduce production costs and make the coal sector profitable. The reduction in employment will be achieved by normal retirements (52,000), early retirements (12,000), and the transfer of employees to other industrial sectors (14,000) (*ibid.*: 17). In order to improve the hard coal sector's debt *vis-à-vis* the social security institution ZUS (*Zakład Ubezpieczeń Społecznych*) the plan suggests writing off interest, to have a repayment holiday of four years of the main overdues, and to split the repayment over a five-year period. It is emphasized that an improvement of the financial situation in the hard coal sector is dependent on decreased production costs. Only two coal mines (Budryk and Bogdanka) are expected to become fully privatized within the next few years. However, there will be a great effort to privatize various buildings and infrastructure, which belong to the mines but which are not directly related to the mining activities (*ibid.*: 44).

[266] For example, the 179 million tonnes of coal produced in 1989 created 70 million tonnes of waste (World Bank, 1993c: 30).

[267] The most common movements are pure vertical movement, differential settlements, horizontal movement, and curvature (*ibid.*: 34).

[268] Personal communications with Leszek Preisner and Tadeusz Pindor, 1996. They claim that a minor "earthquake" occurred in Bytom 1989 in which 243 houses were damaged. Ryszard Fedorowski, a Katowice-based journalist specializing in the coal mining sector, claims that several buildings in the Paderewski housing estate in central Katowice have changed their position because protection pillars have been exploited by a coal mine (personal communication, 1996).

- Air pollution. During the 1980s some 150 million tonnes of coal were annually burned in Poland, causing substantial emissions of sulphur dioxide, particulate matter, and carbon dioxide (World Bank, 1993c: 35). Minor parts (approximately 1 per cent) of the emissions were caused by energy production for the coal mines' own needs.
- Saline waste water discharges (see below for more details).

Additional problems are enhanced levels of radium which appear in the coal beds, and which are pumped out together with the waste water,[269] noise (particularly from the ventilation systems in the mines), and emissions of methane.[270] The price of Polish coal would certainly look different if all of the above-mentioned environmental problems were internalized in the coal price.

The 1980 Statute on the Protection and Shaping of the Environment identifies nine environmental protection obligations for the hard coal sector:

- treatment of waste water;
- introduction of technologies for closed water circulation;
- prevention and elimination of contaminated underground water;
- introduction of modern mining technologies;
- waste management;
- desulphurization;
- action to alleviate surface subsidence;
- recultivation of slag heaps; and
- liquidation of mining facilities in the least damaging way.

The legal development was followed by the creation of a department for environmental protection within the Ministry of Mining and Energy. This ministry was dissolved in the late 1980s. After systemic change the Polish Hard Coal Agency has played an important role in the coordination and development of environmental protection measures.[271]

In addition, there are other pieces of legislation of importance for environmental protection within the hard coal sector, for example, the Water Law, the Mining and Geology Law, the Statute on Land-Use Planning, the Statute on the State Inspectorate for Environmental Protection, and the Building Law (Dulewski, 1996).

An extensive system of environmental charges has been applied *vis-à-vis* the hard

[269] The concentration of radium in the waste water from certain mines reaches 17,000 Bq per square metre water. This should be compared with the Polish norm for radium in drinking water, which is 110 Bq (Węgrzynowska, 1993). A complicating factor is that the half-time for radium is several thousand years. Poborski (personal communication, 1996) argues that no real risk assessment has been carried out. The Main Mining Institute in Katowice has developed a technology for elimination of radium 226 that is presently used in some hard coal mines (NSW, 1996: 9). The issue of radium in the hard coal mines was highlighted in the popular weekly magazine *Wprost* in 1994. In this article, which was entitled "Silesian Chernobyl", the journalist Bartosz Dąbrowski (1994) wrote that the level of radium was above the permissible level in twenty-six mines.

[270] The emission of methane, an aggressive greenhouse gas, was 628,330 tonnes in 1992 (Gawlik and Grzybek, 1995: 372).

[271] Personal communication with Marian Chaber, 1996.

coal sector since the 1980s.[272] Coal mines that violate the emission limits and/or environmental quality standards are obliged to pay fines.

The salination problem

In extracting coal, many mines in Upper Silesia discharge waste water containing high concentrations of saline minerals.[273] All hard coal mines in Upper Silesia discharge underground water to the rivers. Thirty-two coal mines pump out their water to the Wisła basin, thirty-six mines discharge water to the Odra basin. Half of the mines in the former group discharge saline water (1.8-42 g/l), thirty-one mines of the latter group (Chaber, 1994). Of the 400 million cubic metres of water pumped out of the coal mines in Poland 1989, about two thirds was considered to be usable, the remaining third was excessively saline (World Bank, 1993c: 33).

The reason why the problem appears in this region is that the seams of the coalfield lie in predominantly sandstone strata in which waters within this aquifer gradually increase in salinity with increasing depth. In some of the deeper mines the water is three times more saline than seawater. In the 1980s the daily discharge of salt was close to 7,000 tonnes.[274] In 1992 the discharges had declined to the level of 4,800 tonnes (OECD, 1995a: 112). The major part of the saline water (approximately three quarters) is pumped out to the Wisła river; the rest ends up in the Odra river. The salinity of the water in the southern parts of these two rivers is in certain locations higher than in the Baltic Sea.[275]

The highest permissible level of concentration of sulphates in the river water is 0.4 g/l (Karaczun and Indeka, 1996: 80). In the Wisła river the water contains more sulphates than this level to the town of Kazimierz, situated more than 300 kilometres from the polluting coal mines. In dry years the permissible level is sometimes transgressed in the Wisła's river water in the Warszawa region (Chowaniec and Świst, 1994). In Kraków, a city situated some hundred kilometres east of the Katowice region, the concentration of sulphates reaches up to 2.2 g/l (NSW, 1996: 3).[276] Some

---

[272] Besides the environmental charges there are exploitation fees for mining activity (personal communication with Hanna Ogulewicz, 1996).

[273] Saline water is measured as $Cl^- + SO_4^{2-}$. Dry kitchen salt, sodium chloride (NaCl), is neutral. When salt is dissolved in water, negatively and positively charged ions are created. The most common way of expressing salinity equivalents in water is to indicate the content of chloride ions (Cl^-) and sulphate ions (SO_4^{2-}).

[274] For instance, 6,900 tonnes were discharged daily in 1989 (OECD, 1995a: 112).

[275] As a rule, raw water concentrations are measured as total dissolved solids content (TDS) in g/l of water. Based on the degree of salinity, raw water can be classified in the following way: Saline water of low concentration: 0.5-3 g/l TDS; brackish water: 3-20 g/l TDS; seawater: 20-50 g/l TDS, and brine: more than 50 g/l TDS. Oceans have an average concentration of 35 g/l TDS (Ribeiro, 1996: 23). On an average, the salinity of the water in the Baltic Sea is somewhat below 10 g/l. Some of the water which is discharged by the mines Czeczott, Piast, and Ziemowit contains more than 42 g salt per litre (MEPNRaF, 1996).

[276] In November 1993, an automatic monitoring station for saline water in the river Wisła started to operate in Kraków. Measurements made in May 1996 showed that 93.6 per cent of the tests had a level of chloride higher than the permissible norm for industrial use (Regionalny Zarząd, 1996: 1).

280 kilometres of the Odra river have increased concentrations of salinity (Forowicz, 1996).

The concentration of chloride ions in the Wisła river near Piast, Ziemowit and Czeczott, the mines discharging most of the saline water,[277] amount to 5.55 g/l (Komisja do Spraw Ocen Oddziaływania na Środowisko, 1996: 1). In Bieruń Nowy, a small town near these mines, the average concentration is 7.3 g/l (Chowaniec and Świst, 1994).

The negative environmental impact of the salination can be divided into three categories. Firstly, harm to ecosystems and river life. Secondly, increased costs for drinking water treatment downstream. The increased salinity in the rivers should be seen in the light of the fact that Poland is poorly endowed with freshwater resources, and the ratio for available water resources is one of the lowest in Europe (OECD, 1995a: 47). For many urban residents, surface water is the main source of drinking water (*ibid.*: 22), a fact which makes Poland vulnerable to pollution of the surface water. Thirdly, material costs. According to one estimation, the annual economic losses due to corrosion caused by saline water may be as high as $100 million.[278] The productivity costs of these discharges probably account for the largest component of total losses due to water pollution in Poland (0.5-0.8 per cent of GDP) (World Bank, 1994). In Kraków, saline river water results in considerable problems for certain types of industrial activity (OECD, 1995a: 51). A large number of companies in Kraków that are dependent on water from the river Wisła claim that they have to change pipes with increasing frequency due to salination damage.[279] The costs for replacing the pipes have become larger and larger. The saline water in the river Wisła has also been recognized as a problem by the heating sector in Warszawa (OECD, 1995a: 51).[280]

Prognoses have been made over estimated future discharges of saline water. According to one estimation (Rogóż, 1994), the situation by the year 2005 will be the following: water pumped out by the mines will decrease by 14 per cent but the concentration of chloride and sulphate ions will rise by 67 per cent, due to increased average depth of excavation. Hence, in the absence of a radical solution salination of Polish rivers is likely to become an increasingly serious problem.

---

[277] The quality of Polish river water is divided into four classes: Class I: suitable for drinking; Class II: suitable for agriculture and tourism; Class III: suitable for use in industry; and Class IV: not suitable for any purposes. The largest polluters with respect to Class IV water in 1994 were Ziemowit: 933.6 tonnes/day, Piast: 898.0 tonnes/day and Czeczott: 859.5 tonnes/day (personal communication with Hanna Baradziej, 1996).

[278] Personal communication with Zenon Raczkowski, 1996.

[279] A World Bank mission that visited a combined heat and power plant in Kraków reported the following: "They had installed desalination equipment, but this was inadequate to meet the full water requirement. As a result, they estimated that the life of pipework and other equipment that came into contact with water from the Wisła was halved due to corrosion. This caused great inconvenience and substantial losses of energy during the winter because it was difficult to schedule the replacement of pipework during the non-heating period of the year" (World Bank, 1992b: annex A: 8).

[280] A study referred to by ING BH Consultants *et al.* (1994: A-11) estimates that the total damage caused to the environment by saline water discharges has a value of $215 million. This figure has been reached by taking account of additional costs for industrial and municipal waste water treatment, clean water intake from more distant sources, losses in fishing, recreation and tourism losses, corrosion in heating and water systems, decrease in self-purification ability, and value of lost raw materials in discharges.

The main polluter: NSW

The hard coal mines belonging to Nadwiślańska Coal Company (*Nadwiślańska Spółka Węglowa* or NSW) contribute 93.5 per cent of the salt discharges to the Wisła river (NSW, 1996: 3) (Table 7.3)[281] NSW is situated in Tychy near Katowice and was created as a joint stock company on 1 March 1993. All in all there are eight hard coal mines that belong to NSW (Figure 7.1).[282] NSW is the largest of the coal companies in terms of production. The main discharges of NSW originate from Czeczott, Piast, and Ziemowit. These three mines are responsible for 87 per cent of NSW's saline discharges which in 1994 totalled 3,520 tonnes per day of $(Cl^- + SO_4^{2-})$ salts. In other words, three NSW mines provide nearly 80 per cent of the total chloride discharge (ING BH Consultants *et al.*, 1994: A-10).

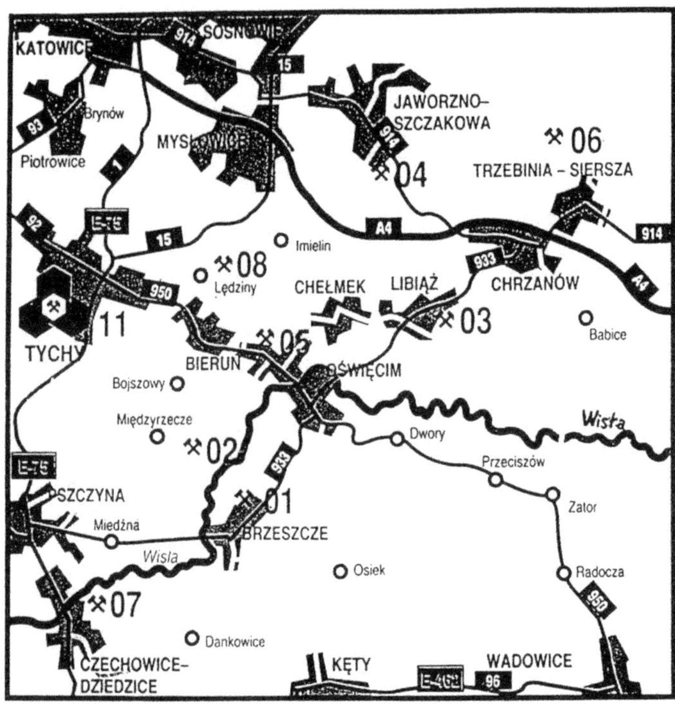

Figure 7.1 Coal Mines in Nadwiślańska Coal Company: 1. Brzeszcze, 2. Czeczott, 3. Janina, 4. Jaworzno, 5. Piast, 6. Siersza, 7. Silesia, 8. Ziemowit
Source: NSW, 1994.

---

[281] In 1995 the discharge of chloride and sulphate ions reached 3,263 tonnes per day. The discharges by the year 2000 are expected to increase to some 4,540 tonnes per day (NSW, 1996: 3).

[282] The following mines belong to NSW: Brzeszcze, Czeczott, Janina, Jaworzno, Piast, Silesia, Siersza, and Ziemowit. The oldest mine is Jaworzno, established in 1792, the youngest one is Czeczott, founded in 1985. In 1993 NSW produced some 25 million tonnes coal which constituted approximately 20 per cent of the national production (NSW, 1994).

Table 7.3 Amount of saline waste water discharges from the coal companies in 1993
($Cl^- + SO_4^{2-}$ tonnes/day)

Coal company	Discharges	Coal company	Discharges
Bytomska	159	Rybnicka	390
Nadwiślańska	3,348	Jastrzębska	270
Gliwicka	456	Other mines	54
Katowicki Holding	366	Total	5,232

Source: Chaber, 1994.

7.3. The policy process

Before 1989

The hard coal sector's salination problems were recognized as early as in the 1960s and 1970s. First and foremost, the problem was signalled by various voivodships situated along the Odra and Wisła rivers, which had become increasingly aware that the salination of the rivers not only was a threat to their main drinking water source but also constituted a problem for industry. A number of voivodships demanded that action be taken by the central government and the authorities in Katowice. The problem was also recognized by the GDR which asked Poland for economic compensation for the economic losses that East German industry had incurred due to the salinated water in the Odra region.[283]

In the 1980s the problem was also highlighted by the NGO community. Its most important statement was made by the PKE at its second general congress in February 1987. The club adopted a resolution that warned that the main Polish rivers were likely to be biologically dead by the year 2000 if no measures against saline discharges were taken. The club demanded a number of measures be taken. Firstly, it advocated coal excavation at layers containing less salty water. Secondly, it proposed the construction of desalination plants for heavily salinated waters (containing more than 42 g/l. Thirdly, the club sharply rejected the idea of diluting saline waste in other regions (PKE, 1987: 55-56).

In 1988 the national environmental programme to the year 2010 identified saline waste water discharges as one of the most urgent environmental problems that needed to be addressed (MOŚiZN, 1988a). It made clear that in order to bring all water of the Wisła and Odra rivers up to Class III quality (i.e. suitable for industrial use) it was necessary not only to build a large number of municipal and industrial waste water treatment plants but also to address the salination problem. The programme warned that discharges were likely to double by the year 2005 due to increased production depths and the start of coal production at new mines with waste water of high salinity. It was the policy-makers' intention to solve the problem by (1) the construction of small desalination plants and (2) the creation of a pipeline system for dosing the most saline waste water from retarding reservoirs to the Odra river in the Koźla region and to the upper and central parts of the Wisła river.[284] Financing was to be achieved via

[283] Personal communication with Danuta Plinta, 1997.

[284] As regards dosing saline water to the Wisła river, the programme mentions three possible places: the Wisła above Kraków, near the San estuary, and in the Włocławek region north of Warszawa

the saline account, a special fund set up within the Environmental Protection Fund fed by fees for discharges of saline waste water, introduced in 1987 (*ibid.*: 34-35). The programme estimated that it would take up to fifteen years to mobilize the appropriate financial resources necessary for solving the salination problem (MOŚiZN, 1988b: 76).

The 1990s

The salination problem seems to have lost much of its prominent position on the public and governmental agendas it enjoyed before systemic change. This trend was already discernible at the environmental table of the Round-table discussions between the socialist government and Solidarity in 1989 where the problem was not mentioned at all. Furthermore, in the national environmental policy of 1991 the Ministry of Environment did not put the salination problem among those problems that it considered needed to be dealt with urgently (MEPNRaF, 1991). Also the environmental movement has attached less priority to the issue in the 1990s. Or to put it bluntly: it has not indicated any interest all. A perusal of the main environmental magazines reveals that NGO interest in the environmental problems of the coal mining sector has been minimal. For example, in the Green Brigades, a national magazine which publishes articles by representatives from a large segment of the Polish environmental organizations, there was not a single article about the coal sector between 1989 and 1995 (*Zielone Brygady* 1989-1995). Likewise, the PKE branch in Katowice has not written a single article about the topic in its bulletin during the past few years. Furthermore, in the 1990s no environmental organization has presented any views about how the problem should be solved. This leads to the conclusion that there was more concern about the problem in the 1980s than in the 1990s.[285] The minimal interest in the salination issue shown by the environmental organizations in the 1990s has implied a change in the agenda-building process. The policy process has become more closed and there is little interaction between the actors within the environmental policy subsystem. There were clearly more interactions between the policy actors in the 1980s than in the 1990s. Since systemic change virtually all policy initiatives related to the problem have originated from the government and the hard coal sector itself.

From what has been said above it follows that the policy process was mainly societal corporatist before systemic change, but has been transformed into a state corporatist policy process in the 1990s.

7.4. Major policy issues

7.4.1. Standards and permits

Standards for permissible level of chlorides and sulphates in receiving water were introduced in 1975. These standards have remained the same since then (Table 7.4). Effluent standards for these substances were introduced in 1991. Czeczott and Piast

---

(MOŚiZN, 1988a: 35).

[285] However, in 1995 the President of Kraków suggested that Czeczott, Piast, and Ziemowit be closed (personal communication with Krzysztof Krogulski, 1996).

have never operated with valid permits for their discharges of saline waste water. Hence, in a strict sense their activities have been illegal from the very outset. Ziemowit was granted a permit in 1974 that expired in 1984.[286]

Table 7.4 Maximum permissible concentrations of pollutants in surface water

Substance	Unit	Water quality classes		
		I	II	III
Chlorides	mg Cl/l	250	300	400
Sulphates	mg SO_4/l	150	200	250

Source: Karaczun and Indeka, 1996: 81.

7.4.2. Enforcement

Fines for discharges causing impermissible concentration of salinity in the receiving waters were introduced in 1975.[287]

The three main polluters of saline water, Czeczott, Piast, and Ziemowit, were all included in PIOŚ' list of the eighty largest polluters in 1990.[288] Two of these mines were obliged to minimize their harmful impacts on the environment before 31 December, 1994, another mine was obliged to do so before 31 December 1996 (Karbownik, 1994).

7.4.3. Fees and financing

As already mentioned in Section 7.3, environmental fees for discharges of saline waste water from the coal mines were introduced in 1987 (Komisja Planowania, 1988: 125).

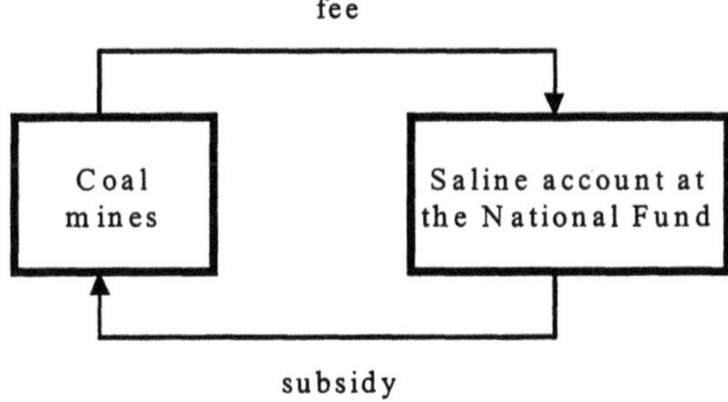

Figure 7.2 Main strategy for addressing the salination problem.

[286] Personal communication with Krzysztof Krogulski, 1998.

[287] Personal communication with Krzysztof Krogulski, 1998.

[288] Another fourteen hard coal mines were included in the regional lists in which 828 companies were included (Karbownik, 1994).

The fees have been earmarked for a so-called "salt account", first located within the Environmental Protection Fund (1987-89) and then, in 1989, moved to the National Fund for Environmental Protection and Water Management. Money that accrued to the account has been reimbursed to the hard coal sector in the form of subsidies for various measures to reduce the discharges of saline waste water (Figure 7.2). Hence, the rationale for setting up this account was to force the coal mining sector to save money for future investments.[289] In other words, the account has functioned as a piggy bank. Before systemic change the fees for saline waste water were kept at a symbolic level. In 1989 the pollution fee for chloride and sulphate ions content was $0.53/tonne. In 1990 the Ministry of Environment, Natural Resources and Forestry increased the rate more than tenfold, and the fee reached $7.89/tonne. Interestingly, there are no indications that the coal sector made any objections to this dramatic increase.[290] After that, the fee level has been periodically adjusted for inflation. In the beginning of 1996 the fee reached a level of $55.15/tonne (Table 7.5).

Table 7.5 Pollution fees for saline water. Chloride and sulphate ions content, rates ($/tonne)

Year	Rate	Year	Rate
1989	0.53	1993	30.31
1990	7.89	1994	39.61
1991	23.62	1995	45.37
1992	40.35	1996	55.15

Source: Śleszyński, 1996b.

There is no doubt that the fee makes a real difference to the financial performance of the coal mines. For example, a representative of NSW claims that, in 1996, 13.6 per cent of the coal price sold by NSW constituted environmental fees; 95 per cent of these fees were related to the discharges of saline water.[291]

7.4.4. Technology

Two major solutions came forward during the 1970s and 1980s, desalination and dilution of the saline water in other regions.

I. *Desalination.*[292] A small-scale pilot plant for desalination by thermal evaporation

[289] Personal communication with Eugenia Koblak, 1996.

[290] Personal communication with Jan Rzymełka, 1996.

[291] Personal communication with Zenon Raczkowski, 1996.

[292] By the end of 1991 there were 8,886 desalination installations in operation over the world. Most of the desalination occurs in Saudi Arabia (24.4 per cent), the USA (15.2 per cent), the Arab Emirate (10.6 per cent), and Kuwait (9.1 per cent) (Motyka and Magdziorz, 1994). In commercial desalination, two main types of technologies have been used extensively throughout the world: thermal processes and membrane processes. The former is based on distillation, where water is evaporated from a saline solution and condensed as fresh water, while the latter makes use of semi-permeable membranes to separate fresh water from the saline water (Ribeiro, 1996: 15). For treatment of mine water with high salinity, desalination by thermal evaporation is considered to be the most effective approach.

was put into operation in 1975 at the Dębieńsko mine (Sikora *et al.*, 1994). The construction of large-scale desalination plants was not considered to be technically or financially feasible. Instead, attention was focused on leading away the waste water from the Katowice region.

II. *Dilution.* For many years many policy-makers argued that the proper solution was to lead the saline water all the way to the Baltic Sea via a large pipeline. This concept was dropped in the mid-1980s. A more realistic idea was to lead the polluted water to places a few hundred kilometres downstream of the river Wisła where the saline water was to be diluted with river water from tributaries to the river Wisła. In dry years the saline water was to be retained in large reservoirs and to be released when the amount of water in the river was considered to be sufficiently high. The technical preparations for this project involved such institutions as the Central Mining Institute in Katowice, the Department for Mining Projects in Kraków and others. The idea was seriously considered until the beginning of the 1990s.[293] There were three reasons why the concept was not realized. First, dilution ceased to be perceived as an effective solution in environmental terms. Second, the concept was thought to be too expensive. Finally, it was discovered that the construction of a long pipeline system would entail complicated negotiations with municipalities about land-use permits (ING BH Consultants *et al.*, 1994: A-38).

After 1989 Poland has been offered foreign assistance to address the problem. For instance, the Japanese company Kawasaki H.I. initiated cooperation with Ziemowit to achieve reduced discharges by way of reversed osmosis and thermal methods for recycling of saline waste water (Chowaniec and Świst, 1994). Simultaneously PHARE, the European Union's aid programme for central and eastern Europe, offered support to a few small projects at the Piast and Czeczott mines.

In 1995 a second desalination plant, Dębieńsko II, was taken into operation to treat saline water from two hard coal mines, Dębieńsko and Budryk. The plant, whose technology is based on reversed osmosis and evaporation, has a capacity to treat some 14,200 m^3 saline waste water per day, which corresponds to a few per cent of the total saline discharges in Upper Silesia (*ibid.*).

In the 1990s various low-cost measures at the coal mines have been investigated and partly implemented. One of the measures that have been realized is the limitation of the supply of saline water to the headings of the mines. This method has been most successfully applied at the Silesia mine where closed mining headings have been sealed and where water has been hampered behind the dams. Similar methods have been tested and used at another thirteen coal mines, including Czeczott, Piast, and Ziemowit.[294]

---

[293] Personal communication with Jan Kazior, 1996.

[294] A number of other small-scale measures have been proposed and partly realized. The most important ones are: (1) Reduction of waste water discharges through actions at the mine. The amount of water varies with the exposed coal seam areas. Concentration of production onto fewer and more productive faces may reduce waste water by 10-15 per cent. (2) Production at deep levels with the most saline water may be reduced (ING BH Consultants *et al.*, 1994: 3-9). The coal mines have tried to focus mining activity above the 500 metres level, where the salt concentration of the water is lower than in strata situated below the 500 metres level (*ibid.*: B-11). (3) The deposition of saline water in headings together with fly ash and postflotation waste. This method is used by thirteen mines to treat some 3,000 m^3 saline water per day. (4) The management of medium saline waste water at the mechanical processing of the coal. This method, relying on filling up silt-water circulation at the plants for mechanical processing of the coal, has made possible the management of some 5,400 m^3 saline waste water per day at fourteen coal mines (e.g. Czeczott). (5) Reinjection of saline waste water into

At the beginning of the 1990s, the Ministry of Environment and NSW agreed upon a plan for the construction of a large-scale desalination plant for the treatment of saline waste water originating from Czeczott, Piast, and Ziemowit. In connection with this decision NSW established Ekosol, a limited liability company, which was given the following tasks with respect to saline water management (Ekosol, 1993):

- to coordinate engineering in the domain of management of saline water by means of utilization, geological engineering, and hydrotechnical methods;
- to optimize the saline water utilization processes in the domain of energetic and chemical technologies and quality guarantees of the products; and
- to act as a substitute investor in designing and realization of technological installations on behalf of the mines in NSW.

The large-scale thermal desalination plant that was agreed to be built[295] should be able to treat 32,680 m^3 per day of the most saline waters (ING BH Consultants *et al.*, 1994). According to the concept, saline water will be carried away via pipelines[296] and treated at a desalination plant which will be built on unoccupied land belonging to the "Oświęcim" Chemical Plant (Komisja do Spraw Ocen Oddziaływania na Środowisko, 1996: 1). An international competitive tender for the realization of the project was won by a consortium consisting of one German and one Polish company, Balcke Dürr AG and Energoprojekt Katowice SA, respectively (NSW, 1996: 6). The former was given responsibility for the design and construction of the desalination plant while the latter company became in charge of the pipelines and other kind of infrastructure.

The total cost of the desalination plant, pipelines and additional steam and power is estimated at 1 billion PLN (Karkoszka, 1996) or $452 million (ING BH Consultants *et al.*, 1994: 5-7).[297] Some 60 per cent of the saline waste water discharges from Nadwiślańska Coal Company will be eliminated thanks to the new investment.[298] The salinity of the Wisła river is expected to be reduced by at least 40 per cent, perhaps even more (Ficek and Ficek, 1996).[299]

---

underground strata. The idea is simple and cheap but technically unproven (*ibid.*: 5-1). (6) Geological methods for limiting the discharge of waste water after it is pumped out of the coal mine. The following technologies have been considered: recirculation, deep and shallow compression, and draining of the coal mine without discharge of waste water (Karbownik, 1995).

[295] The desalination plant is partly based on experiences from the two desalination plants at the Dębieńsko mine and various feasibility studies made by foreign companies such as the Japanese Consulting Institute, Organisation et Environment (France) and others (Chowaniec and Świst, 1994).

[296] Altogether 65 km of pipelines through eight municipalities and towns will be built (Karkoszka, 1996). Positive decisions about the pipeline have to be taken first at the local level and then at the level of the voivodship authorities (ING BH Consultants *et al.*, 1994: 5-11).

[297] The construction of the desalination plant should therefore be considered as one of the largest investments connected with environmental protection in Poland in the 1990s.

[298] Personal communication with Jan Kazior, 1996.

[299] When the whole plant starts to operate it is expected to each year produce 1,200,000 tonnes of evaporated salt together with 400,000 tonnes of other chemicals. Under the condition that the evaporated salt (sodium chloride) produced at the desalination plant will reach a purity of 99.8 per cent it could be sold to the chemical industry (Karkoszka, 1996). For some of the products which will be produced in smaller quantities (iodine and bromine and others) the propects are very good. The income

The desalination plant is expected to be a profitable endeavour for NSW. The financial viability of the investment is an effect of the avoided fees and fines. Without these instruments the internal rate of return ends up at 8 per cent. If the avoided costs of fees and fines are included, the financial viability of the plant increases dramatically and the internal rate of return reaches the level of 23.1 per cent (*ibid.*: 5-7). The desalination plant thus has the potential of becoming a win-win investment.

7.5. Implementation

Since 1992 the National Fund has received approximately 5-10 per cent of the fees that the hard coal mines were obliged to pay for their discharges of saline water. The collection rate of the fees for saline waste water discharges has been considerably lower than for other pollution charges (Table 7.6). This had implied that small amounts of money have accrued to the saline account (Table 7.7). In 1996 the debt of the unpaid fees had accumulated to 700 million PLN, corresponding to some $300 million.[300] Table 7.8 shows that the unpaid fees to the National Fund constituted the largest part of NSW's total liabilities in 1994. There is some evidence that the government led by Hanna Suchocka in 1992-93 secretly exempted the hard coal sector from the obligation to pay its pollution fees.[301]

Table 7.6 Collection rate of environmental fees (%)

Year	*Saline water*	Sulphur dioxide	Nitrogen dioxide	Waste water[302]	Waste disposal	Ground water withdrawal
1990	*119*	99	94	108	107	105
1991	*62*	86	85	69	63	77
1992	*6*	86	80	50	86	83
1993	*5*	150	84	55	68	75
1994	*12*	97	96	45	78	96

Source: Śleszyński, 1996b.

expected from the sales of the chemical by-products has been estimated at $100 million. The creation of a new salt supply may put some existing Polish salt mines out of business and hence create unemployment (ING BH Consultants *et al.*, 1994).

[300] Personal communication with Hanna Ogulewicz, 1996.

[301] Personal communication with Jerzy Wertz, 1997.

[302] Includes: biological oxygen demand, chemical oxygen demand, suspended solids, heavy metals, and saline water.

Table 7.7 The National Fund's revenues of fees for saline waste water
(million US dollars)

Year	Amount	Year	Amount
1990	2.02	1993	2.05
1991	38.27	1994	14.02
1992	5.19		

Source: Śleszyński, 1996b.

Table 7.8 Liabilities of NSW in mid-1994
(million US dollars)

Type of debt	Amount
Suppliers	34.7
State	42.5
Taxes	22.7
Social security	49.3
Credit and loans	74.0
National Environmental Fund	
Fees	154.0
Fines	31.0

Source: ING BH Consultants *et al.*, 1994: 3-7.

In July 1996 the Ministry of Environment, Natural Resources, and Forestry and the Ministry of Industry and Trade began negotiations about the conditions for the repayment of the unpaid environmental fees. The position of the Ministry of Environment was that the coal mines should be allowed to postpone payment of the fees for a few years and to repay the debt, including the interest rates, when the financial situation had improved. Under the condition that the coal mines undertake investments in desalination measures, the Ministry of Environment stated that it is open to investigate the possibility to fully or partly forget the interest rate.[303] Also the Ministry of Industry and Trade was in favour of exempting the mines from the debt. However, it also wanted to exempt the mines from paying fees during the next few years.[304] In December 1996 the parties reached a compromise, which implied that the fee for saline waste water discharges was reduced by 50 per cent (*Trybuna Górnicza*, 1996).

It is obvious that PIOŚ has been a toothless tiger *vis-à-vis* the largest polluters of saline waste water. The dilemma is well summarized by Zbigniew Kamiński, who has been in charge of the list of the eighty largest polluters at PIOŚ: "Realistically it is not possible to achieve effective enforcement. As long as the sector is in a very bad financial situation, compliance with environmental regulations will be problematic".[305]

[303] Personal communication with Eugenia Koblak, 1996.

[304] Personal communication with Marian Chaber, 1996.

[305] Personal communication, 1996.

PIOŚ has been less influential in the hard coal sector than in other sectors. This can partly be attributed to the fact that the environmental inspectorate is only responsible for enforcement of environmental protection measures above the surface, not inside the coal mines. There is a special mining law that states that inside the coal mines the Highest Mining Authority (*Wyższy Urząd Górniczy* or WUG) is responsible for *all* kinds of inspections, including those which concern protection of the environment. WUG issues operational permits for the coal mines and is the main body for supervision at the coal mines. Basically all decisions which concern the operation of a coal mine have to be agreed with a Regional Mining Authority (*Okręgowy Urząd Górniczy* or OUG).[306] A coal mine may only be closed on agreement with an OUG (Dulewski, 1996). The consequence of this is that saline water only becomes a concern for PIOŚ when the saline water has been pumped out of the coal mines. Also the voivodship authorities influence is limited as regards activities taking place underground; the environmental authorities cannot intervene in order to demand changes in the technology. This restriction is defined in mining law.[307] The enforcement failure should also be seen in the light of certain features that make the hard coal mines unique compared to other major polluters.

- Water must be pumped out of the coal mines continuously otherwise floods will occur underground.
- A coal mine is a dangerous work place. When conflicts between environmental protection and working security arise, the latter always has priority.[308]
- Many coal mines are hydrologically interconnected. If one mine is closed and another one in the immediate vicinity is not, water from the closed mine is likely to flow into the one that is still in operation. Every decision to change the drainage system of one coal mine requires careful analysis of the potential hazard for neighbouring coal mines (Darski *et al.*, 1995: 5).[309]

These three examples show that environmental protection is more complicated in the hard coal sector than in other industrial sectors.

Another factor impeding closures of Piast and Ziemowit is that these mines are assumed to be competitive in the future and have a long economic life. Ziemowit is the largest hard coal mine in Europe (with a capacity of 30,000 t/d) and Piast is the second largest coal mine in Poland. Their reserves are large, the geological conditions

[306] There are fourteen of those in Poland of which six are located in Upper Silesia.

[307] Personal communication with Wojciech Beblo, 1996. Beblo, who was the director of the voivodship's environmental office in Katowice between 1991 and 1995, explains how he perceived the situation: "The voivodship's authorities could not intervene in order to change technology in the coal mining sector. In other sectors that was possible. Normally an industry had to agree on a technology with the environmental office. In the mining sector this was not the case because of the legal structure. The coal mines did not become part of my office's repair programme for the heavy industry in the region. I knew about this problem beforehand so the failure was expected".

[308] The pumping out of water is the most obvious example. Another conflict is the ventilation of the mines, which causes a serious noise problem, both for the miners and the people living in the vicinity of the coal mines (Dulewski, 1996).

[309] An investigation of forty hard coal mines showed that all but three (Gliwice, Budryk, and Ziemowit) are hydrologically connected with other coal mines (Darski *et al.*, 1995).

are good, and they are modern in terms of infrastructure. They also belong to the lowest cost per tonne coal producers in Poland (ING BH Consultants *et al.*, 1994: A-34-45).[310]

7.6. Advocacy coalitions and their belief systems

The main actors which engaged themselves in the salination problem in the 1980s were the following: the Ministry of Environmental Protection and Natural Resources, the *Sejm* Committee for Environmental Protection and Natural Resources, PIOŚ, the Department for Environmental Protection in Katowice voivodship, the Environmental Protection Fund (from 1989: the National Fund for Environmental Protection and Water Management), the environmental protection departments at the Central Mining Institute, the Ministry of Industry, the Ministry of Mining and Energy, the Polish Academy of Sciences, the PKE, and a number of voivodships along the Wisła and Odra rivers. Most of these actors were also active on the national environmental policy level and hence were members of the Environmental Protection Advocacy Coalition, EPAC, in the 1980s (see Chapter 4). The key elements in the EPAC's belief system with respect to the salination problem were that it was a priority problem and that the most proper policy instrument was a fee for discharges of waste water earmarked for a saline account. It was assumed that the problem could and should be solved without closures of any hard coal mines. The environmental authorities together with the mines advocated dilution of saline water while the PKE and the most affected voivodships were in favour of desalination.

As shown in Chapter 5, the EPAC was dissolved in the early 1990s. The actors affiliated with the EAC (the Efficiency Advocacy Coalition) and the GAC (the Green Advocacy Coalition) have taken no active interest in the salination issue. The main actors are representatives of the MAC (the Mainstream Advocacy Coalition), the hard coal sector (NSW, Ekosol, PARGWK, the Central Mining Institute etc.) and various research institutes. What actually can be observed is a "marriage" between the MAC and the coal sector or something that could be labelled the Miners' Welfare Coalition. The actors centred around the salination issue share a number of beliefs.

Policy core beliefs

The proper relationship between environment and economy. The actors hold a number of beliefs about the hard coal sector's role in the Polish economy which shows that they consider economic concerns to have priority over environmental ones. Firstly, the salination problem cannot be effectively addressed until the economic performance of the hard coal sector improves. The solution of the salination problem is linked to the financial performance of the hard coal sector. This in turn is related to the possibilities of reconstructing the sector. Most actors recognize that the economic situation is unlikely to improve in the absence of a reconstruction of the hard coal sector. Such reconstruction would imply a limited amount of closures of some of the most notorious loss-makers, a smooth reduction in manpower, improved management, and the introduction of more modern technologies.

[310] Czeczott must be closed before the year 2010 because of resource depletion. Six other hard coal mines will be closed for the same reason (Radetzki, 1997: 39).

Secondly, mass unemployment in the hard coal sector can trigger political instability. The miners are seen as a powerful group that has the potential of destabilizing the political situation in Poland.[311] Several events have occurred in the 1990s that support this way of reasoning. For example, in December 1992 strikes broke out in mines over wages and proposals to restructure the coal industry (Millard, 1994: 100) and in May 1995 over 10,000 miners marched on Warszawa demanding governmental action to protect their jobs (Simpson, 1996). With these weapons the miners are assumed to have the power to bring down any government.

Thirdly, domestic coal is a guarantee for Poland's energy security. As shown in Chapter 3, Poland is almost exclusively dependent on coal as an energy source. In the future, Upper Silesia may become the single region in Poland where hard coal is produced. In recent years virtually all political groupings have advanced the argument that Poland needs a strong domestic coal sector in order to guarantee the energy security of the country. There are several reasons for this. First, by relying on coal Poland reduces its dependency on imported fuels. Most importantly, the present and planned gas and oil deliveries from Russia are not considered to be entirely safe because of the unstable economic and political situation in that country. Second, Poland is a flat land where the potential for water power is limited. Third, Poland has no nuclear power. The nuclear programme which started in the 1980s was stopped in 1990. According to Polish energy analysts, nuclear power may be considered by the year 2010. There are unofficial energy policy papers that forecast some 6,000 MW of installed nuclear power capacity by that time (Kramer, 1995: 98).

Whose welfare counts? It is clearly the welfare of the miners and their families that is the main concern for the policy-makers dealing with the salination issue. The social cost for mass unemployment in the hard coal sector is considered intolerable. Radical employment changes in the hard coal sector would have far-reaching social implications because the hard coal sector is the main employer in Upper Silesia. Out of the four million people in the former Katowice voivodship, approximately 300,000 are coal miners. If the miners' families are included, it appears that one million people are directly dependent on the hard coal sector. If the auxiliary sectors that support the mining sector are also taken into account, the number of people who are dependent on the coal mines reaches some 1.5 million.[312] The absence of alternative employment and adequate social benefits, means that the miners and their trade unions are unlikely to agree on radical manpower reductions.[313]

The seriousness of the problem. The salination problem was recognized as a priority issue in the 1980s. In the 1990s the problem has not achieved this status.

The basic cause of the problem. The basic cause of the salination problem is seen as the lack of money to address the problem, just as in the 1980s. In the 1980s lack of technology appeared as another barrier.

Proper scope of governmental vs. market activity. In the 1990s it has been widely believed that a precondition for a far-reaching reconstruction of the hard coal sector is

---

[311] In Jerzy Śleszyński's words (personal communication, 1996), "nobody would commit political suicide by closing the mines". Marian Chaber (personal communication, 1996) claims that "if there were coal mine closures due to environmental reasons, the miners would go with their pick-axes to the authorities".

[312] Personal communication with Andrzej Klasik, 1996.

[313] More than 140,000 people have left the mines since 1990.

the development of alternative employment opportunities for the miners. Hence the assumption that the government should play an active role in creating new work places. It is generally argued, both inside and outside the hard coal sector, that the absence of alternative work places is related to the *laissez-faire* approach to industrial policy, which the government had adhered to in the beginning of the Polish transition.[314]

The proper distribution of authority has not been discussed by the policy actors.

The choice of policy instrument and method of financing. The general assumption here is that the tools being used until now to address the salination problem (the saline account at the national fund collecting fees for saline waste water discharges which are recirculated back to mines) are appropriate. This holds for actors both in the 1980s and the 1990s.

The ability of society to solve the problem. In both the 1980s and the 1990s the discourse has been characterized by technological optimism.

Who should participate? This question has not been directly addressed by the policy actors but the implicit assumption has been, both in the 1980s and the 1990s, that the issue of salination could be delegated to experts.

Secondary aspects beliefs

Main technological solution. In the 1980s the EPAC advocated dilution and desalination. In the 1990s it is only desalination that has been put forward as an appropriate technological solution.

The belief systems of the actors dealing with the salination problem in the 1980s and the 1990s are depicted in Table 7.9.

Policy-oriented learning

The main learning event that can be observed in this case study is that the idea to dilute saline waste water in the Odra and Wisła rivers has been dropped and replaced with the conviction that desalination is the most appropriate technical solution. In the 1990s policy-oriented learning has been impeded by (1) the low degree of public participation and debate in the policy process and (2) the general perception that the hard coal sector should continue to play an important role in Poland's economy.

---

[314] A new approach to regional development in Upper Silesia has been elaborated within the framework of the so-called Regional Contract for the Katowice Voivodship, a ten-year public-private partnership (including the trade unions) which aims at promoting new work places via industrial reconstruction and various initiatives for the regional development, and accelerate industrial reconstruction (Simpson, 1996). The Regional Contract includes the establishment of a Fund for Upper Silesia which aims, *inter alia*, to support people who want to start up a new economic activity. The contract also promotes the establishment of economic free zones and the development of infrastructure (Ministerstwo Przemysłu i Handlu, 1996: 18-19).

Table 7.9 Main beliefs about the salination problem

	1980s **Environmental Protection Advocacy Coalition**	**1990s** **Mainstream Advocacy Coalition and Miner's Welfare Coalition**
Policy core		
Environment vs. economy	Environmental improvements can be achieved without detrimental effects for the miners and the hard coal sector.	The salination problem cannot be addressed until the economic performance of the hard coal sector improves substantially. Large-scale closures of mines should be avoided because (1) mass unemployment in the hard coal sector can trigger political instability; and (2) the coal is needed for Poland's energy security.
Whose welfare counts?	Miners *and* all persons negatively affected by the salination of the rivers.	Miners. The social consequences of large-scale closures of the mines are intolerable as long as there are no alternative employment opportunities available for the miners.
The seriousness of the problem	A priority problem.	*Not* a priority problem.
Basic cause of the problem	Lack of money and technology to treat saline waste water.	Lack of money to treat saline waste water.
Proper scope of government vs. market activity	Not discussed.	The government should play an active role in industrial policy.
Proper distribution of authority	Not discussed.	Not discussed.
Choice of policy instrument	See method of financing.	See method of financing.
Method of financing	Fee for discharges of saline waste water and earmarked subsidies from the "saline account".	Fee for discharges of saline waste water and earmarked subsidies from the "saline account".
Ability of society to solve the problem	Technological optimism.	Technological optimism.
Who should participate?	Not discussed.	Not discussed.
Secondary aspects		
Main technological solution	Dilution and desalination.	Desalination.

7.7. Summary and conclusions about policy change

The hard coal sector in southern Poland discharges large amounts of saline waste water to the Odra and the Wisła rivers. Environmental fees for saline waste water discharges are supposed to be paid by the polluters. The fees are collected at an "saline account" at the National Fund for Environmental Protection and Water Management and earmarked for desalination investments. This model has remained virtually unchanged since 1987 but has constantly failed to mobilize sufficient financial means required for solving the problem. The problem remains unsolved. The salination of the main rivers represents perhaps the largest implementation failure in Polish environmental policy in the 1990s. The policy-makers dealing with the salination problem face political, social, and economic constraints that make implementation extremely problematic.

This problem was high on the environmental agenda in the 1980s. However, in the 1990s much less attention has been paid to this problem. The actors involved in addressing the problem in the 1990s are representatives of the MAC and the hard coal sector. The GAC and the EAC have abstained from participation.

Closure of the most polluting mines is considered as an intolerable option because of the social consequences of such a measure. Moreover, the hard coal sector is equated with national self-sufficiency and hence the salination issue has a link to Poland's capacity to make a successful transition to an independent statehood.

The salination issue is an intriguing example of overlapping subsystems, in this case between the environmental policy subsystem and the mining subsystem. The unequal power positions between these two subsystems can explain the dominating position of the mining subsystem. In the 1990s it has been the hard coal sector that sets the agenda and defines the alternatives.

The salinity problem is a pure national concern. In contrast to most other environmental issues in Poland there are no international conventions or EU directives that urge Poland to take action.[315] Arguably, the absence of international driving forces for addressing the problem has contributed to the implementation failure.

[315] The Environmental Action Programme for Central and Eastern Europe recommends countries in transition to "invest in mitigating discharges of saline water from mines" (MEPNRaF, 1995). But besides this brief recommendation, there has been no international pressure for Poland to solve the salination problem. The international commission against pollution of the Odra has no immediate plans to address the salination problem (personal communication with Alfred Dubicki, 1998).

PART IV

CONCLUSIONS

8. CONCLUSIONS ABOUT POLICY CHANGE

So far, the book has dealt with the development of environmental policy in Poland before and after systemic change in the chapters on national policy in the 1980s (Chapter 4), national policy in the 1990s (Chapter 5), air pollution from stationary sources (Chapter 6), and saline waste water discharges from the hard coal sector (Chapter 7). Environmental policy has been addressed from four perspectives: characteristics of the policy process, policy content, policy implementation and policy impacts, and policy belief systems. Hence, the first subquestion that was formulated in Chapter 2 – What environmental policy has developed before and after systemic change? – has been answered.

The task of this chapter is to answer the remaining research subquestions: To what extent was change radical or incremental? What have been the main driving forces for policy change and continuity? It also answers the key research question: To what extent has systemic change in Poland affected environmental policy in the 1990s?

Section 8.1 investigates whether environmental policy changed in a radical way or in an incremental way. In Section 8.2 conclusions are drawn about the driving forces for policy change. Section 8.3 answers the key research question. Section 8.4 contains a short evaluation of the ACF. Section 8.5 briefly summarizes the chapter.

8.1. Change – incremental or radical?

Change in the policy process

1980s. In terms of agenda-building models, an important change occurred in the 1980s compared to the 1970s: issues arose not only within the government but also outside the government. In other words, the agenda-building process of the 1980s can be explained by both the inside initiative model and the outside initiative model (Cobb *et al.*, 1976). This can be illustrated by three developments. First, in the 1980s it is possible to observe more and more incorporation of experts, such as engineers and scientists affiliated with the Polish Academy of Sciences, in the policy-making process. This trend was partly related to the emergence of an increasingly complex industrial society with demands on more specialized division of labour. Second, the first independent environmental organization in Poland – the Polish Ecological Club – was formed in 1980. In the second part of the 1980s a large number of new independent environmental organizations appeared. The activities of the NGO-community became an increasingly significant factor driving the pace of environmental policy reform in the 1980s. The aluminium plant in Skawina was closed in 1981 as the result of a PKE campaign. The creation of the Ministry of Environmental Protection and Natural Resources was partly a PKE achievement. PKE's advocacy of eco-development (sustainable development) in the middle of the decade was translated to official policy at the Round-table between the government and Solidarity in 1989. PKE and several other environmental organizations had representatives in advisory boards to the Ministry of Environment.

However, there was, just as in the 1970s, little direct interaction between the environmental sector and the economic sectors. The main reason for this was that environmental policy could simply be ignored by the economic actors in the absence of rule of law.[316] They had no incentives to interact with the environmental policy-makers in any stage of the environmental policy-making process.

Throughout the 1980s, martial law in 1982-83 included, the environmentalists were allowed to promote their belief system with relative impunity. Environmental policy-making was one of the few policy areas that allowed for public participation in Poland during the 1980s. How was this possible? A first explanation is that the government wanted to defuse western pressure on Poland's human rights record by diverting attention to environmental issues. According to this line of thought, the political leadership perceived environmental protection as a suitable way of "seeking contact" with the international community in order to lessen Poland's political and economic isolation in the wake of martial law. Likewise, environmental protection may have been perceived as a useful tool for the government to gain legitimacy at home. As Cobb and Elder (1983: 15) have concluded, "legitimacy (...) is always conditional and must be earned and sustained if the system is to retain popular confidence". A second explanation is that Poland's leadership used environmental issues as a way of channelling public discontent to a politically relatively safe area. In other words, environmental protection was a tool in a political risk management strategy. Lastly, it is also possible that environmental policy deliberately was made a test case for political pluralism.

One of the reasons why the environmentalists were so popular in the 1980s was that some of them challenged the socialist system. The environmental movement of the 1980s clearly had a double mission: on the one hand it aimed at improving the environmental protection potential of the existing system, on the other hand it was a Trojan Horse for the political opposition.

1990s. Three major trends are discernible in the 1990s as far as the policy actors are concerned. First, the NGO community has been growing rapidly but is more marginalized in the broader political and societal context than in the 1980s. Second, the interactions between the environmental policy subsystem and the economic subsystems have increased substantially in the 1990s. Large segments of industry, notably the power sector, have become interested in environmental policy. This is related to the fact that environmental policy has begun to make a difference to their financial performance and the insight that environmental protection measures may be beneficial for Polish companies' competitiveness on western markets. Third, the role of the scientific sector is declining, partly as a result of its difficult financial situation.

Environmental policy appears to be an a-political policy field in the 1990s. It is noticeable that only a few political parties have bothered to develop explicit views on environmental policy. It is experts rather than politicians who deal with the task of environmental protection. As Millard (1998: 153) correctly observes, "(...) the political component of environmental policy-making seems to have been largely absent. Policy is debated in narrow circles, academics, and consultants of varied ilk".

---

[316] However, in a few instances, environmental policy made a difference for the economic sectors: a few closures of polluting plants occurred and some developmental plans were delayed or postponed.

In terms of agenda-building models, the 1990s show a mixture of the inside initiative model and the outside initiative model. The Ministry of Environment's attempt in 1990 to mobilize society for the task of improving the environmental performance of the eighty largest polluters corresponds well with the mobilization model. However, no other events fit with this model after systemic change.

The air protection policy reflects most of the trends that can be observed in national policy-making. A major difference between national policy-making and the air protection policy in the 1990s is that environmental NGOs have been much more active in the former than in the latter. This is related to the fact that the environmental movement perceives the emissions from stationary sources to be a problem which, by and large, is under control.

The policy process in the salination issue has had corporatist features both in the 1980s and 1990s. In the 1990s a collusion between environmental management and the hard coal sector is observable. A relatively closed circle of environmental protection experts and representatives of the hard coal sector dominate the policy process. It is the hard coal sector that has set the agenda; the saline account could be seen as a non-decision made by the policy-makers to let the coal sector decide itself if, when, and how to address its discharges of saline waste water.[317]

In conclusion, the Polish environmental policy experience appears to show an interesting mixture of both a pluralist, incremental way of policy-making, as hypothesized in the ACF, and the corporatist models as presented by Ziegler. The most noteworthy findings are (1) the shift from state corporatism in the 1970s to societal corporatism in the 1980s, both on the national and sectoral levels, and (2) the absence of pluralist forms of interactions in relation to the salination issue in the 1990s (Table 8.1).

Table 8.1 Policy-making styles identified at national and sectoral level

Period	National policy	Air pollution	Saline waste water
1970s	State corporatism	State corporatism	State corporatism
1980s	Societal corporatism	Societal corporatism	Societal corporatism
1990s	Pluralism	Pluralism	State corporatism

Change in the policy content

At the time of systemic change, "there was an option of whether to develop new policy framework or whether to adjust elements of the old regulatory system to evolving circumstances" (Śleszyński, 1996b: 126). This book points to the second alternative. Thus, the notion of a clean break with the past followed by an immediate and dramatic change is not applicable.

Change in policy goals. In terms of quantitative environmental goals there is a high degree of continuity between the 1980s and 1990s. The general goal of a sustainable development was emphasized for the first time in a governmental policy paper just before systemic change in 1989. The goals set by international conventions have played much more important roles in the 1990s than in the 1980s.

Policy instruments. The current pollution control system in Poland rests on three main pillars:

[317] See Andersson, 1998 and Andersson and Hisschemöller, 1997.

1. A financing system based on environmental fees that are paid by the polluters, collected by the environmental authorities, and used by environmental funds on national, regional, and local levels for various environmental protection investments.
2. National environmental quality standards.
3. Pollution permits issued by the voivodships' departments for environmental protection.

By and large, this system was already fixed by the 1980 Statute on the Protection and Shaping of the Environment. In the 1990s we witness an adjustment of instruments shaped in the socialist system to the new economic and political order. This conclusion is supported by the developments on the sectoral level. The instrumentation for dealing with the air pollution from stationary sources and the salination problem has been the same in the 1980s and 1990s.

Table 8.2 Main policy instruments in the pollution control system: 1975, 1985, and 1995

	1975	1985	1995
Enforcement	Fines	• Fines • Closures (made by voivodships' environmental protection departments) • Inspections (carried out by PIOŚ)[a]	PIOŚ in charge of: • Closures • Fines • Inspections • Monitoring
Fees	Fees for: • Water use and water pollution	Fees for: • Water use and water pollution • Air pollution • Waste disposal	Fees for: • Water use and water pollution • Air pollution • Waste disposal
Funds	Water Management Fund	• Water Management Fund • Environmental Protection Fund	• National Fund for Environmental Protection and Water Management • Voivodship Environmental Funds • Municipal Environmental Funds • Ecofund
Standards	National Environmental Quality Standards	National Environmental Quality Standards	National Environmental Quality Standards
Permits	Issued by the voivodships' environmental protection departments	Issued by the voivodships' environmental protection departments	Issued by the voivodships' environmental protection departments

[a] State Inspectorate for Environmental Protection

Economic instruments have played a central role in the pollution control system, both before and after the demise of socialism. Fines have been in use since the 1960s and water fees since 1975. The 1980 environmental framework act substantially expanded the role of the fees and fines. In the 1990s there has been a sharp increase in the fee rates and a development and refinement of the environmental fund system. The main pillars of the pollution control system before and after systemic change are presented in Table 8.2.

Institutional setting. The institutional framework for environmental policy has remained relatively stable over the studied period. The backbone of the present environmental protection administration was built in the 1970s and the 1980s. Environmental protection departments in the voivodships were already in place in the 1970s. Environmental issues became part of the cabinet formation in the beginning of the 1970s, which is early in an international perspective. The Ministry of Environmental Protection and Natural Resources was formed in 1985. Some changes have occurred within this Ministry after systemic change. Most important was that the Ministry became in charge of Forestry in 1989.

PIOŚ was established in the early 1980s. Its role has been upgraded in the 1990s since it has become in charge of monitoring of environmental quality and has been given the right to impose fines and to interrupt activities in breach of environmental law.

The financing system of the 1990s rests on initiatives taken in the 1980s. The environmental protection fund and the water management fund of the 1980s have been merged into the National Fund for Environmental Protection and Water Management in 1989. On the voivodship level, independent environmental funds have been established, also based on existing structures.

There are also new financing institutions in the environmental protection subsystem that have appeared in the 1990s: the Ecofund, the Bank for Environmental Protection, and the municipal environmental funds. However, they account for a relatively small share of the total environmental protection expenditures.

Since the demise of the centralized bureaucracy of the socialist system, more and more emphasis is placed on decisions at the smaller administrative levels of society. This general tendency is also discernible in the environmental protection subsystem. Seven Regional Water Management Boards were established in 1991 to manage and develop water use within the river basins, and the municipalities have been given a greater role in the land-use planning process. (Appendices 1 and 2 show the structure of environmental protection administration in Poland in 1985 and in 1995.)

Change in implementation and policy impacts

The agenda-building approach may be used to explain the phenomenon that Poland's environmental policy in the 1980s was, by and large, a success as regards policy formulation and adoption (at least compared to preceding decades) but a failure with respect to implementation. As Cobb and Elder (1983: 152) have pointed out, formal agenda access is no guarantee for success:

> Having a place on the formal agenda does not necessarily mean that action will be taken. Often, actions will be deliberately delayed. All that is required for formal agenda status is that the leaders know that there is a problem that requires governmental actions and make some type of response, even if it is only official recognition of the problem.

To a certain extent environmental policy in Poland in the 1980s had the properties of what Cobb and Elder (1983) call a pseudo-agenda item.

Systemic change has, generally speaking, entailed a radical improvement of Poland's environmental performance. Between 1989 and 1995 the pollution intensity of the Polish economy fell substantially. The fall in air pollution has implied that Poland, as shown in Chapter 6, has been able to comply with the first sulphur protocol of the Convention on Long-Range Transboundary Air Pollution. In the mid-1990s the

Polish government declared that it considers four areas to be in ecological hazard, compared to the twenty-seven areas that were designated in the 1980s.

Enforcement, performed by the State Inspectorate for Environmental Protection, has improved in the 1990s. Only two industrial plants were closed in the 1980s while more than 100 enterprises have been closed, partly or completely, in the 1990s. Enforcement has become more effective due to the fact that the state no longer fulfils the dual functions of an owner and a regulator of the economy. Enforcement is also facilitated by the imposition of rule of law.

> (...) the Inspectorate would not be having a significant impact (...) were it not for systemic reforms, most notably the institution of a constitutional *Rechtsstaat* (literally "law state") in Poland. (...) during the socialist era, environmental statutes, like all laws, were mere policy instruments that the Party/state simply disregarded (...) whenever they proved inconvenient. That is no longer the case (Cole, 1998: 206).

The financing mechanisms have been strengthened in the 1990s. The rates of the fees and fines were raised by more than ten times in real terms in the early 1990s. The annual expenditures of the environmental funds increased by almost twenty times between 1990 and 1993. Poland's environmental expenditures in relation to the GDP reached 1 per cent in 1991 which was a level that had never been reached before. A direct result of increased revenues is that the number of waste water treatment plants taken into operation each year increased almost tenfold between 1980 and 1996. Likewise, the capacity to treat gaseous pollution increased more than tenfold between 1985 and 1996.

The introduction of a market economy entails increased budget discipline among the polluters. In contrast to the 1980s, they now have financial incentives to limit emissions and conserve resources (Cole, 1998: 249).

In the 1990s a number of new initiatives have been launched to increase the integration of environmental concerns into the sectoral policies. However, the commitment to environmental protection and sustainable development has not been sufficiently internalized by the major ministries.

The hard coal sector represents a major exception with respect to policy implementation. The discharges of saline waste water remain, as shown in Chapter 7, an unsolved problem due to a number of economic, political, and social constraints.

Change in the belief systems

Technological optimism and a trust in socialism's capacity to deal successfully with environmental problems characterized the belief system of the environmental policy-makers in the 1970s. Interestingly, the political establishment of the 1970s saw the promotion of environmental policy as an opportunity for gaining international political legitimacy. The wish to present Poland as a modern, open, and pragmatic state and a reliable business partner to the West was a first departure from the orthodox socialist belief system. This departure became much more articulated in the 1980s. In almost all respects, environmental problems were framed differently in the 1980s than in the 1970s. First of all, the economic and political context was included in the debate and questions began to be asked about the social order in which the environmental degradation occurred. The fact that socialism was no longer seen as a panacea to the environmental policy problems represented a change in the deep core of the belief system. Policy principles such as prevention, precaution, and integration were advocated. The interest in market-oriented instruments increased substantially.

There was a widespread recognition of the seriousness of the environmental situation. During the 1980s the positions of the environmental authorities and the environmental movement converged more and more. The ultimate proof of this process was the 1989 Round-table, which showed great consensus on almost every environmental issue. In the 1990s the dominating belief system has been that of the Mainstream Advocacy Coalition (MAC). It contains elements of the dominating belief systems of the 1970s and the 1980s. For example, the MAC's views on the basic cause of the environmental problems and the ability of society to solve these problems are similar to the views of the 1970s on these matters. The MAC's advocacy of sustainable development and a decentralization of the environmental management system together with its positive approach to an increased public participation show affinity with ideas that were held by the EPAC in the 1980s. The dominating belief system of the 1970s, the 1980s, and 1990s are compared in Table 8.3.

A radical change has occurred in the 1990s with respect to belief systems. Three new Advocacy Coalitions have emerged – the Efficiency Advocacy Coalition (EAC), the Green Advocacy Coalition (GAC), and the Mainstream Advocacy Coalition (MAC) - largely in response to the new challenges faced by the new systemic framework. The most important difference between these coalitions' belief systems is on the deep core level: in contrast to the EAC and the MAC, the GAC considers man to be a part of nature. On the policy core level the EAC believes that adequate policies may avoid conflicts between environmental protection and economic growth. The MAC regards trade-offs between these two goals to be unavoidable while the GAC emphasizes that environmental goals are more important than other goals, including economic ones. The EAC considers poorly designed policies to be a major cause of environmental problems, while the GAC considers consumption to be the most important cause of these problems. In the MAC's view, it is the lack of money and technology which is causing problems. With respect to policy instruments the EAC is a strong advocate of emission trading and other forms of market-based incentives, the GAC is in favour of a green tax reform, and the MAC has a preference for Negotiated Compliance Schedules. There is a striking difference in the GAC's and the MAC's perceptions about the ability of society to solve environmental problems: the GAC is a technological pessimist while the MAC is an optimist in this regard. On the secondary aspects level the GAC is worried by the environmental consequences of Poland becoming a member of the European Union. The MAC does not see any problems with Poland's entry to the European Union, rather the opposite. The EAC accepts Poland's entry to the European Union but argues that EU environmental policy is neglecting cost-effectiveness.

The mere fact that the Polish transition has "produced" an efficiency-oriented coalition is quite remarkable. It is even more remarkable that the Polish power industry is a part of the EAC. The concern behind its advocacy of marketable permits is a desire to improve its economic performance. It can be concluded that the market economy has "(...) served to disaggregate industrial interests, so that Polish industry no longer presents a united front on (or against) environmental policy" (Cole, 1998: 228).

The collusion of the environmental administration and the hard coal sector has implied that some alternatives, such as closures of the most polluting mines, have not been seriously considered. It is assumed that hard coal should continue to play a central role in the Polish economy. Attention is focused on finding financial resources for a large-scale desalination plant.

Table 8.3 A comparison between the dominating environmental belief systems in the 1970s, 1980s, and 1990s

	1970s	1980s	1990s
		Dominating coalition	
	No advocacy coalition in place.	The Environmental Protection Advocacy Coalition (EPAC).	Mainstream Advocacy Coalition (MAC).
		Deep core	
	Socialism seen as infallible, a panacea.	Socialism not seen as fallible, no panacea.	Man dominates nature. Positive approach to market economy.
		Policy core	
Environment vs. economy	First get rich, then protect the environment. Environmental policy is a policy area within its own right.	"Eco-development". Pollution is a developmental barrier. Environmental protection is profitable.	Formal commitment to sustainable development. Trade-offs between environmental and economic concerns.
Whose welfare counts?	Formally: the working class.	The citizens.	No clear-cut preferences but sectoral privileges not excluded.
The seriousness of the problem	Bad but not very bad.	Ecological catastrophe.	The environmental situation is improving.
Basic cause of the problem	Lack of obedience, technology, and money.	Production most important policy goal. Lack of prevention.	Lack of money and technology.
Proper scope of government vs. market activity	Not discussed.	Not discussed.	No clear-cut preferences.
Proper distribution of authority	Central authorities are the key actors.	Regionalization and decentralization.	Decentralization.
Choice of policy instrument	National ambient quality standards.	National ambient quality standards. Fees and fines. Land-use planning. EIA.	Present tools + Negotiated Compliance Schedules.
Method of financing	State subsidies.	Environmental funds. PPP in the long run.	Environmental funds.
Ability of society to solve the problems	More of the same to solve the problems. Technological optimism.	Conceptual changes needed. Technology not the only solution.	Technological optimism.
Who should participate?	Experts.	Public participation very important.	Public participation important.
		Secondary aspects	
Institutional aspects	Political loyalty most important.	Professionalism most important.	Professionalism most important.
Priority problems	Water management and nature protection.	Public health, air protection, water management, waste water, nature protection.	Air protection, waste water, waste.

Conclusion

The four policy categories - process, content, implementation, and belief systems - have not been equally changed in the investigated period. In terms of policy process there were radical changes in the early 1980s, both with respect to agenda-setting models and forms of political interactions. In the 1990s there is no radical change in the agenda-setting but a new form of political interaction, pluralism, has emerged. The change in policy content was radical in the 1980s but incremental in the 1990s. The first part of the 1980s was a period of radical changes with respect to the choice of policy instruments and institutional arrangements. In the second part of the 1980s the goals of environmental policy changed radically. As far as implementation is concerned, change was incremental in the 1980s and radical in the 1990s. The most radical changes in implementation occurred in the first few years after systemic change. The belief systems of the policy actors have been radically changed both in the 1980s and the 1990s. The most intense changes occurred in the Solidarity period in 1980-81 (the end of the belief in socialism) and in 1992-93 (the emergence of three new advocacy coalitions).

8.2. What have been the main driving forces for policy change and continuity?

The impact of external events and stable parameters other than systemic change

The most important driving forces for environmental policy change in Poland have been the Stockholm Conference in 1972, the Solidarity period in 1980-81, the Chernobyl catastrophe in 1986, and the approximation of Polish law to EU legislation in the 1990s. The Stockholm Conference stimulated the elevation of environmental protection to ministerial level and gave the impetus for the elaboration of the environmental framework law, the Statute on the Protection and Shaping of the Environment. The importance of the Solidarity period was that an independent environmental organization, the Polish Ecological Club, emerged and new patterns of interactions developed rapidly. Chernobyl not only produced strong anti-nuclear sentiments (that would contribute to the decision to cancel the nuclear programme in Poland) but also led to a general activation of the environmental movement. In recent years the most important driving force for continued policy-reform has been Poland's will to harmonize its environmental legislation with the one in the European Union.

Three of the above-mentioned external events occurred *before* systemic change. It is also interesting to note that the impact of the West has been largest in the 1970s and the 1990s. These are decades when the economic and political relationships between Poland and the West have been redefined. In connection with this, environmental policy has been directly linked to the foreign policy agenda. In contrast, in the 1980s it was political developments inside Poland, together with Chernobyl, which served as the most crucial driving forces for policy change.

The salination issue represents a stark contrast to most environmental issues in Poland in that it is not affected by the harmonization of Polish environmental law to EU directives. This is attributable to the fact that the salination problem does not appear commonly in Europe and that the saline discharges only affect Polish territory. Since Poland has no international obligations with respect to saline waste water this issue has not become a priority issue.

The lack of pressure from the EU is not the only constraint facing the policy-makers addressing the salination problem. Also the threat of unemployment, Poland's coal dependence, the social structure inUpper Silesia, and the financial situation of the hard coal sector are substantial obstacles.

With respect to the air protection policy, it is clear that the integration process with the EU constitutes a constraint for the EAC since emission trading is not in use within the EU. Poland's dependence on coal is a constraint for all actors concerned about achieving an improved air quality situation. The Polish population is relatively poor, which means that the potential for further energy price increases – which provide incentives for energy use reductions – is low.

The role of policy-oriented learning

By and large, policy-oriented learning in the 1970s was a matter of imitation of international trends. It is reasonable to assume that the elevation of environmental protection to ministerial level, the elaboration of the environmental framework law, and the inclusion of environmental protection into the 1976 constitution would not have occurred without similar developments in the West. However, the introduction of economic instruments in Poland in the 1960s and 1970s was inspired by developments in the Soviet Union rather than in the West. Already in the mid-1960s, Soviet economists had become interested in enhancing the effectiveness of environmental protection by the incorporation of economic incentives.

After 1989, lesson-drawing from abroad has continued to be an important factor in the policy reformation process. For example, the creation of seven Regional Water Management Boards in 1991 was modelled on the French water management system and the Environmental Impact Assessment Commission of 1990 was modelled on the Dutch EIA Committee (Bowman and Hunter, 1992). The 1991 Nature Protection Act is based on the IUCN's World Conservation Strategy (Jendrośka, 1996: 379). I should emphasize that I consider the hitherto approximation of Polish environmental law to EU legislation to be a matter of "policy-taking" rather than something that can be called lesson-drawing or learning.

The international community influences the three advocacy coalitions in the Polish environmental policy subsystem in the 1990s in different ways. The EAC is inspired by the US environmental policy approach while the MAC has an orientation towards the approaches advocated by the EU. The GAC is open to influences from western environmental NGOs (such as Greenpeace and Friends of the Earth International) as well as the green parties.

Since environmental policy was a single-coalition subsystem in the latter part of 1980s, it can be concluded that policy-oriented learning in this period took place *within* one coalition. This implied that virtually every policy change reflected an element of learning and, *vice versa*, many elements of learning were translated to policy. In other words, policy-oriented learning was a very important factor for policy change in the 1980s. This study identified more instrumental learning than conceptual learning in the 1980s. To recall, instrumental learning relates to changes in the secondary aspects of a belief system while conceptual learning concerns changes in the policy core of a belief system (see Section 2.4).

In the 1990s three coalitions have emerged. Still, most of the learning processes concern secondary aspects of the beliefs system. Learning within belief systems is much more frequent than learning across belief systems. The connection between policy-oriented learning and policy change is less evident in the 1990s than in the 1980s.

The most sophisticated learning processes in the 1990s have occurred in relation to the air protection policy. The learning that has occurred within the EAC's belief system is worth a special mention. The EAC has thoroughly analysed the weaknesses of the policy instruments presently in use and subsequently put forward a number of innovative suggestions to improve the efficiency and effectiveness of the air protection policy. The EAC's learning processes have been facilitated by the fact that many of its members are highly skilled researchers (some of them with international reputation) with international experience and a high ability to interact with other policy actors and to create networks. It could be argued that the EAC has more incentives to learn than the MAC and the GAC because it is the smallest coalition in the environmental policy subsystem. As a minority coalition the EAC has come to the conclusion that its only chance to have success, that is, to translate its beliefs to concrete policy, is to outlearn the other coalitions. The EAC has managed to create a limited commitment to emission trading within the MAC and among some of the realists within the GAC. This is the only example of conceptual learning across belief systems that this study has identified.

The air policy case study showed that the GAC refrains from involving itself in detailed and technical policy debates. The coalition has obviously not developed the capacity for doing this. Also, many members of the GAC do not have an interest in becoming involved in technical debates. Instead, they focus their attention to issues of more general nature.

Compared to national environmental policy, there has been an exceptionally low degree of policy-oriented learning in the salination issue in the 1980s and the 1990s. The instrumentation (fees for saline water collected at a saline account within the National Fund for Environmental Protection and Water Management) has been unchanged and unchallenged since 1987 despite its conspicuous failure to address the problem. The major learning event that has taken place during the past decade is that the idea of dilution of saline waste water has been dropped and that desalination has appeared as the most proper technological solution. This is a change in the secondary aspects of the belief system. How should the low dynamics of the learning processes be explained? One reason is, in my view, that the issue has involved a narrower range of opinions and fewer participants than other issues in the environmental policy subsystem.

Table 8.4 Policy-oriented learning found at national and sectoral level

National policy	Air pollution	Salination
Instrumental learning within a belief system		
• National ambient quality standards (1960s) • Permits issued by voivodships (1960s) • Water pollution fees (1974) • Water management fund (1974) • Expanded role of pollution fees (1980) • Closure of industry possible (1980) • Environmental protection fund (1980) • Fines for pollution must be paid from company profits (1983) EPAC (1985-91) • EIA introduced (1985) • Need to create the National Fund for Environmental Protection and Water Management recognized (1988) • Regionalization and decentralization of environmental protection advocated (1988) • Polluter Pays Principle advocated (1988) • Debt-for-nature swap advocated (1989) • Regional Water Management Boards proposed (1989) • Bank for Environmental Protection proposed (1989) • The right to close a factory transferred from the voivodship level to PIOS (1991) • New role of the state in env. prot. (1991) MAC: • Independent voivodship funds (1993) • Local environmental funds (1993) • Enforcement *vis-à-vis* industry should be more flexible. NCS advocated (1996) GAC: • Green tax reform advocated (1996) • It is not enough to protest. Well-prepared proposals are required (1991-92) • Improved coordination of the activities is required (1993)	• Air quality standards (1965) • Air pollution fees introduced (1980) • Fines four times higher than fees (1980) • Closure possible (1980) EPAC (1985-91) • Air pollution is a developmental barrier (1985) • Fines ten times higher than the fees (1987) • Stricter ambient quality standards (1990) • National emission standards for combustion processes (1990) • Fee increase with more than 1,000 per cent in real terms (1990-91)	• Fines for saline discharges (1975) • Standards for receiving water (1975) • Dilution proposed (1970s) EPAC (since 1985) • Fee for saline waste water introduced (1987) • Saline account created (1987) • Effluent standards for saline waste water introduced (1990) • Sharp increase in the fee rate (1990-91) • Desalination replaces dilution as the main technical solution for solving the problem (early 1990s) • 50 per cent reduction of the fee rate (1996)
Instrumental learning across belief system		
MAC accepts increased transparency of the National Fund's disbursement procedures (1993)		
Conceptual learning within a belief system		
EPAC: • Eco-development advocated by PKE (1985) and Ministry of Env. (1988) GAC: • Consumption is the largest problem (1992)	EPAC: • Emission trading advocated (1989-91)	
Conceptual learning across a belief system		
	• GAC partly open to emission trading, (1995) • MAC open to limited use of emission trading (1995)	

The role of competition between advocacy coalitions

Within the ACF there is a hypothesis stating that "the policy core attributes of a governmental programme are unlikely to be significantly revised as long as the subsystem advocacy coalition which instituted the programme remains in power" (Sabatier and Jenkins-Smith, 1999). This hypothesis can help to explain the high degree of continuity in Poland's environmental policy between the 1980s and the 1990s. In the early 1990s the Ministry of Environment was led by ministers - Maciej Nowicki of EAC and Stefan Kozłowski of the GAC - who advocated radical changes. However, they were in office too short periods to be able to realize their ideas. Since mid-1992 the environmental ministers have been linked to the MAC. All of them have had a cautious approach to reforms such as the legalization of emission trading.

The salination issue raises the following question: why have two of the advocacy coalitions in the environmental subsystem, the GAC and the EAC, voluntarily abstained from participating in the debate about the salination problem? The GAC's silence in the 1990s may be related to the strong anti-nuclear sentiments in the environmental movement. The greens may suspect that a deep reduction in the coal production can renew the government's interest in nuclear power. The GAC has made the choice that is more important to keep Poland free from nuclear power than to solve the environmental problems of the hard coal sector. A second explanation is that the GAC does not see that the hard coal sector can be directly linked to the consumer society, which is the object of its strongest criticism. The linkage between the consumer society and environmental degradation is much more evident in issues such as transport, nature protection, and municipal waste.

The EAC's disinclination to involve itself in the salination problem has, in my view, another explanation. It is difficult for this coalition to promote its belief system in relation to a problem that cannot be solved by the application of market-based instruments. In other words, the salination problem does not have a structure that corresponds to the solutions that the EAC normally is advocating.

8.3. The environment and systemic change

The key research question of this study can now be answered in the following way. The content of environmental policy in Poland was *not* fundamentally changed by systemic change in 1989. The high degree of continuity in environmental policy between the 1980s and 1990s should be seen in the light of the following factors. First, the political change in Poland in 1989 was peaceful and non-confrontational. This may explain why so much of the socialist structures and processes survived systemic change (Millard, 1994: 47). Second, at the time of systemic change the belief was widely held that the old approaches could work in a new political and economic setting. The existing rules were seen as essentially good but not enforced. Many policy-makers share Śleszyński's (1996a: 126) view that there was "a legacy of sometimes fairly acceptable environmental policy and large sets of economic instruments, which the previous system never really tried to implement". Third, at the beginning of the 1990s most of the political attention was focused on the most fundamental goals of the transition – that is, the establishment of a new economic and political framework. There were too limited time and resources for devoting full attention to other tasks. Fourth, the opposition and environmental movement had some input in the environmental policy-making process already before the demise of socialism. Fifth, radical policy change was partly blocked by inertia and lack of various capabilities in the environmental administration: "while political regimes and economic systems can collapse and vanish, the people with their knowledge, habits, practices, affiliations, informal networks, and organizations remain" (Pickel, 1993: 147). It was not possible to replace old personnel and to develop new professional skills as fast as an "ecological shock therapy" would have required.[318] Jendrośka (1996) even argues that the reform process intentionally was gradual in order to avoid the problem of establishing requirements that the administration would be unable to effectively administer. Poland's experience is not unique in this respect because institutions typically change incrementally, something that the Nobel laureate Douglass North has pointed out:

How and why they change incrementally and why even discontinuous changes (such as revolution and conquest) are never completely discontinuous are a result of the imbeddedness of informal constraints in societies. Although formal rules may change overnight as the result of political or judicial decisions, informal constraints embodied in customs, traditions, and codes of conduct are much more impervious to deliberate policies. These cultural constraints not only connect the past with the present and future, but provide us with a key to explaining the path of historical change (North, 1996: 6).

Hence, the policy-making process under the transition period is affected by informal constraints that come from "socially transmitted information and are a part of the heritage that we call culture" (*ibid.*: 37). There is a cultural filter that tends to provide continuity. The tension between the altered formal rules (the new systemic framework) and the informal constraints produces outcomes that have importance for the way environmental policy changes. I would like to point to three elements in Polish culture which constitute contraints for environmental policy reform. There is, as shown in Chapter 3, a perceived division between the rulers and the ruled. Most Poles see politics as a game, which is beyond the sphere of their influence. This has led to an inclination to disregard reforms of *any* kind. Furthermore, weak executive power and fragile political coalitions have been common phenomena in Polish politics in the 1990s. One implication of this has been that environmental policy reforms have been interrupted. Lastly, environmental consciousness is underdeveloped in Poland. For instance, there are hardly any elements in the Polish national identity which are *directly* linked to nature.

Sixth, environmental policy reform in a country in transition is a complicated process from the point of view of coordination.

Even the most radical changes over particular issues, in order to be operational, must somehow be related to the overall institutional framework. In this respect, contrary to the expectations of many, the transition period is not a paradise for policy-makers and law-makers. In fact, from the technical side, policy-making and legal drafting becomes a nightmare when, in a relatively short period, all policies and all laws are redesigned (Jendrośka, 1996: 382).

---

[318] This is a general phenomenon in eastern Europe. In Jancar-Webster's (1995: 59) words: "(...) the relative influence or ability to act of any given agency continues to be based on pre-transition bureaucratic politics. In most cases, environmental personnel have changed little from the communist period, except perhaps in top positions. Many have been working in the environmental bureaucracy for at least a decade and are thoroughly acquainted with national problems and priorities. Likewise, many are simply continuing or expanding programmes began under the communist regime. There has been remarkably little change in thinking about environmental problems".

In other words, the new environmental policies have to be elaborated at the same time as the rules for making policy constantly change.

In terms of implementation systemic change had a major impact. The effectiveness of environmental policy increased after the socialist system had been abolished. When rule of law was introduced it implied that important actors, such as the state administration and the polluters, became subordinated to the legal system. This increased the possibilities to enforce environmental policy. The end of state ownership and central planning implied that companies had to begin to pay attention to costs. Thereby they could not afford to neglect environmental fees and fines. The end of Poland's isolated position in the world economy has contributed to the structural changes of the Polish economy which have occurred in the 1990s. These changes have generally speaking, been beneficial for the environment.

Systemic change has produced radical changes in the belief systems of the policy actors. The appearance of new advocacy coalitions in the 1990s is directly linked to the transition to democracy and market economy.

With respect to the policy processes the picture is mixed. In terms of agenda-setting the most important changes occurred during the Solidarity period in 1980-81. However, pluralist patterns of interactions in the policy-making process were not discernible until the market economy and the rule of law were in place. In connection with this, the economic actors (industry included) realized that they had a direct financial interest in becoming involved in the environmental policy-making process.

This study does not point to the conclusion that the demise of socialism in Poland was a result of environmental activism in the 1980s. However, the relative freedom, which was granted to the actors in the environmental policy subsystem and the policy-makers' reliance on economic instruments, indicated that the ruling elite in Poland was not disinterested in political pluralism and market solutions. This in turn showed that the socialist belief system in Poland was in disintegration. Socialism had become a "malleable alloy of old, set doctrines and new, more down-to-earth ideas enforced by real life" (Kornai, 1992: 417-418). This disintegration of official ideology was also observable in several other policy subsystems in Poland and in 1989 it came to undermine the foundation of the political and economic system.

8.4. A brief evaluation of the ACF

The ACF is a useful framework for explaining environmental policy change in Poland, a country in transition from a system based on Soviet style socialism to a market-based democracy. The ACF's assumptions about what matters in the policy-making process are no less valid in Poland than in any other country that has been the topic of earlier studies based on the ACF.

According to the ACF, studying policy change requires the investigation of a period of ten years or more. This study confirmed the usefulness of taking a long-term perspective on policy change. In Poland the most important changes occurred in the early 1980s (the policy process), the late 1980s (policy adoption), and the 1990s (policy implementation).

This study showed that the activities of a single policy entrepreneur can explain the emergence of the Efficiency Advocacy Coalition in the Polish environmental policy subsystem in the 1990s. The ACF could be strengthened and further refined by taking into account the Policy Entrepreneur Model. This model could complement the ACF in order to increase knowledge about developments on the microlevel of the policy process.

The ACF should be more explicit about the fact that relatively stable parameters and external events not only constitute constraints and resources for advocacy coalitions in their attempts to achieve changes in laws and various policy documents, but are also of fundamental importance for the implementation stage. Implementing environmental policy in a socialist setting is indeed different from implementing environmental policy in a capitalist one. A strong recommendation from this study is that implementation be brought to the fore of the ACF.[319]

The ACF implicitly assumes that policy actors have a particular goal in mind and calculate and weigh the risks of not attaining their objective against the cost of compromise. However, the modern world is dynamic and the definition of many problems is in a state of flux. For instance, environmental politics has become the art of taking good decisions on insufficient evidence. The ACF should acknowledge the implications of this by paying more attention to the interactions between parties who engage themselves in finding or structuring policy problems. This can improve the quality of the ACF.

8.5. Summary and final remarks

Environmental policy in Poland in the 1990s is a mix of change and continuity. With respect to implementation and the structure of the policy actors' belief systems profound changes have occurred in the 1990s compared to the 1980s. However, systemic change in Poland did not lead to a major change in the content of environmental policy. The environmental policy-makers have employed a cautious approach to new instruments and institutions during the transition period.

It is noteworthy that important changes in the policy process in the investigated period occurred *before* systemic change. For example, radical changes in the agenda-building patterns took place in connection with the Solidarity period in 1980-81. This event provided impetus for the creation of an environmental civil society. As hypothesized by the ACF, events external to the environmental policy subsystem have constituted important driving forces for policy change both before and after 1989. Poland's environmental policy experience points to the conclusion that the transition in eastern Europe operates faster at the systemic level than at the subsystem level.

[319] The same recommendation is made by Grin and Hoppe (1997).

REFERENCES

Aghevli, B. B., E. Borensztein, and T. van der Willigen (1992). *Stabilization and Structural Reform in the Czech and Slovak Federal Republic: First Stage*. Washington, DC: International Monetary Fund, Occasional Paper 92.

Ågren, C. (1988). "Poland. Towards Betterment". *Acid News*, No. 2, May.

Aguilar Fernandez, S. (1994). "Convergence in Environmental Policy? The Resilience of National Designs in Spain and Germany". *Journal of Public Policy*, No. 1.

Alberski, R. (1996). *Polityka ochrony środowiska w Polsce*. Wrocław: Towarzystwo Naukowe Prawa Ochrony Środowiska.

Alcamo, J. (ed.) (1992). *Coping with Crisis in Eastern Europe's Environment*. Lancaster, UK: Parthenon.

Almarcha Barbado, A. (ed.) (1993). *Spain and EC Membership Evaluated*. London: Pinter.

Anderson, G. (1994). "Addressing Mobile Sources of Air Pollution in Poland: Emerging Trends and Policy Options". *European Environment*, Vol. 4, Part 6, December, pp. 9-14.

Anderson, G. and B. Fiedor (1997). "Environmental Charges in Poland". *Harvard Institute for International Development. Environment Discussion Paper*. No. 16.

Anderson, G. and T. Żylicz (1997). "Maximizing Leverage of Poland's Environmental Funds: Too Generous or Too Restrictive". *Harvard Institute for International Development. Environment Discussion Paper*. No. 30.

Andersson, M. (1990). *Miljöproblem och miljöpolitik i Polen*. Solna, Sweden: Naturvårdsverket.

Andersson, M. (1995). "Change and Continuity in Poland's Environmental Policy Since 1980". Paper presented at the European Environmental Conference, University of Nottingham, 11-12 September.

Andersson, M. (1996). "Debt-for-environment swap. The establishment of Poland as a responsible global citizen". Paper written within the framework of the NAKE course "Economics of transition", Amsterdam University.

Andersson, M. (ed.) (1997). *From Intention to Action. Implementing Sustainable Development*. Uppsala, Sweden: The Baltic University Programme.

Andersson, M. (1998). "Nine Lessons from Poland". *International Environmental Affairs*, Vol. 10, No. 1, Winter, pp. 3-7.

Andersson, M., S. Gullgren, and N. Sundström (1993). *Efter murens fall – om västs reaktioner på sammanbrottet i öst*. Göteborg, Sweden: Nerenius & Santerus Förlag.

Andersson, M. and M. Hisschemöller (1997). "The Usefulness and Limitations of the Advocacy Coalition Framework: the Case of National Environmental Policy-Making and the Hard Coal Sector in Poland". Paper presented at the conference of Transformation Processes in Eastern Europe arranged by the ESR of the Netherlands Organization for Scientific Research (NWO), Amsterdam, the Netherlands, 6-7 March.

Arp, H., H. Cesar, and J. Parker (1991). "Environmental Policy Cooperation between Eastern and Western European Countries". *Environment*, Vol. 33, No. 6, pp. 44-45.

Aura (1979a). "Dla nas i następnych pokoleń". No. 6.

Aura (1979b). "W poczuciu odniesionych sukcesów i potrzeby dalszych działań". No. 7.

Aura (1980a). "Zdrowie – pierwsza potrzeba człowieka". No. 9.

Aura (1980b). "Nie zagubić ochrony środowiska". No. 10.
Aura (1980c). "Coraz większe kłopoty ze ściekami". No. 11.
Aura (1980d). "Nowe akty prawne". No. 12.
Aura (1981a). "Zdrowie: esencja naszych działań". No. 6.
Aura (1981b). "Posłanie Polskiego Klubu Ekologicznego do IX Nadzwyczajnego Zjazdy PZPR". No. 9.
Aura (1981c). "Solidarne społeczeństwo a środowisko naturalne". No. 11.
Aura (1982a). "Z dyskusji przedkongresowej". No. 9.
Aura (1982b). "Komitet Inżynerii Środowiska PAN". No. 12.
Aura (1983a). "10 lat AURY". No. 1.
Aura (1983b). "Klub dziennikarzy przy Lidze Ochrony Przyrody". No. 3.
Aura (1984). "Rada Społeczno-Gospodarcza o Gospodarce Przestrzennej". No. 5.
Aura (1985). "Narada w Urzędzie Rady Ministrów". No. 4
Ayres, E. (1995). "Europe's Environment". *World Watch*, Vol. 8, No. 6, November, December.
Bachratz, P. and M. S. Baratz (1970). *Power and Poverty. Theory and Practice.* London and Oxford: Oxford University Press.
Balcerowicz, L. (1995). *Socialism, Capitalism, Transformation.* Budapest: Central European University Press.
Barde, J-P (1995). "Environmental Policy and Policy Instruments". In: H. Folmer, H. Landis Gabel, and H. Opschoor (eds.), *Principles of Environmental and Resource Economics. A Guide for Students and Decision-Makers.* Aldershot, UK: Edward Elgar.
Baszkiewicz, D. (1985). "Trzy z minusem". *Przegląd Techniczny.* No. 45.
Batt, J. (1991). *East Central Europe from Reform to Transformation.* London: Pinter.
Baumgartl, B. (1997). *Transition and Sustainability: Actors and Interests in Eastern European Environmental Policies.* London: Kluwer Law International.
Bieroń, W. and E. Garścia (1981). "Przesłanki Decyzji". *Aura*, No. 3.
Biuletyn Informacyjny (1995). IX Ogólnopolskie Spotkanie Ruchów Ekologicznych. Kolumna '95.
Biuro Studiów i Ekspertyz Kancelarii Sejmu (BSE) (1995). *Od Lucerny do Sofii.* Tom 8. Warszawa.
Blanchard, O., R. Dornsbush, P. Krugman, R. Layard, and L. Summers (1991). *Reform in Eastern Europe.* Cambridge, MA and London: MIT Press.
Blanchard, O. J., K. A. Froot, and J. D. Sachs (eds.) (1994). *The Transition in Eastern Europe: Country Studies.* Vol. 1. Chicago, IL: University of Chicago Press.
Bluffstone, R. A. and T. Panayotou (1995). "Optimal Environmental Liability Policy for Central and Western Europe". *Harvard Institute for International Development. Environmental Discussion Paper.* No. 7.
Bochniarz, Z. and R. Bolan (eds.) (1990). *Designing Institutions for Sustainable Development: A New Challenge for Poland.* Białystok Technical University, Białystok.
Bochniarz, Z. and D. Toft (1995). "Free Trade and the Environment in Central Europe". *European Environment*, Vol. 5, pp. 52-57.
Boehmer-Christiansen, S. and J. Skea (1991). *Acid Politics: Environmental and Energy Policies in Britain and Germany.* New York and London: Belhaven Press.
Böhringer, C. (1996). "Fossil Fuel Subsidies and Environmental Constraints: A General Equilibrium Analysis of German Coal Subsidies and Carbon". *Environmental and Resource Economics*, No. 8, pp. 331-349.

Bojarski, W. (1988). "Alternatywa Energetyczna". *Tygodnik Powszechny*. No. 6.
Bowman, M. and D. Hunter (1992). "Environmental Reforms in Post-Communist Central Europe: from High Hopes to Hard Reality". *Michigan Journal of International Law*, Vol. 13, No. 921, Summer, pp. 923-980.
Braybrooke, D. and C. E. Lindblom (1963). *A Strategy of Decision: Policy Evaluation as a Social Process*. New York: Free Press.
Bressers, J., Th., A. and D. Huiteman (1996). "De invloed van beliedsvorming op de vormgeving van economische beleidsinstrumenten: Politics as usual". *Beleidswetenschap*, No. 2, pp. 154-180.
Broad, W. J. (1993). "Sovjet ljög om kärnavfall". *Svenska Dagbladet*, 28/4.
Budnikowski, A. (1992). "Wspózależność ekologiczna Polski i krajów wspólnoty Europiejskiej". *Ekonomia i Środowisko*, No. 1, pp. 75-87.
Budnikowski, A., M. Lubiński, and M. Rusiński (1988). *Polska a problemy ekologiczne świata*. Warszawa: Instytut Wydawniczy Związków Zawodowych.
Business Central Europe (1997). "Of Hype and Halos. A Survey of Poland". February.
Campbell, R. W. (1991). *The Socialist Economies in Transition: A Primer on Semi-Reformed Systems*. Bloomington, IN: Indiana University Press.
Caprio, G. Jr and R. Levine (1994). "Reforming Finance in Transitional Socialist Economies". *The World Bank Research Observer*, Vol. 9, No. 1, January.
Carter, F. W. (1998). "Political Transformations and the Environment". In: H. Van der Wusten (ed.), *Conference Proceedings. Transformation Processes in Eastern Europe. Part IV. Political Transformation and the Environment*. The Hague: ESR.
Carter, F. W. and D. Turnock (eds.) (1993). *Environmental Problems in Eastern Europe*. London: Routledge.
Chaber, M. (1994) "Wody kopalniane problemem górnictwa węgla kamiennego". *Wiadomości Górnicze*, No. 9.
Chang, H-J and R. Rowthorn (1995). "Role of the State in Economic Change: Entrepreneurship and Conflict Management". In: H-J Chang and R. Rowthorn (eds.), *The Role of the State in Economic Change*. Oxford: Clarendon Press.
Chmielewski, T. J. (1988). *Czy przetrwamy*? Warszawa: Nasza Księgarnia.
Chowaniec, J. and K. Świst (1994). "Rozwiązania technologiczne i ramowy program zagospodarowania zasolonych wód dołowych z kopalń Nadwiślańskiej Spółki Węglowej SA". *Wiadomości Górnicze*, No. 9.
Ciesielski, R. (1988). "Dynamiczne wpływy górnicze na powierznię terenu". *Aura*, No. 5.
Cirtautas, A. M. (1997). *The Polish Solidarity Movement. Revolution, Democracy, and Natural Rights*. London: Routledge.
Clague, C. and G. Rausser (eds.) (1993). *The Emergence of Market Economies in Eastern Europe*. Cambridge, MA: Blackwell.
Claudon, M. P. and T. L. Gutner (eds.) (1992). *Comrades Go Private: Strategies for Eastern European Privatization*. New York and London: New York University Press.
Cobb, R. W. and C. D. Elder (1983). *Participation in American Politics. The Dynamics of Agenda-Building*. London: Johns Hopkins University Press.
Cobb, R., J. K. Ross, and M. H. Ross (1976). "Agenda Building as a Political Process". *The American Political Science Review*, Vol. 70, pp. 126-138.
Cohen, M., J. G. March, and J. P. Olsen (1972). "A Garbage Can Model of Organizational Choice". *Administrative Science Quarterly*, No. 17, pp. 1-25.

Cole, D. H. (1995a). "An Outline History of Environment Law and Administration in Poland". *Hastings International and Comparative Law Review*, Vol. 18, No. 2, pp. 297-358.
Cole, D. H. (1995b). "Poland's Progress. Environmental Protection in a Period of Transition". *The Parker Journal of East European Law*, Columbia University. Vol. 2, No. 3, pp. 279-319.
Cole, D. H. (1995c). "Environmental Protection and Economic Growth: Lessons from Socialist Europe". In: R. P. Malloy and C. K. Braun (eds.), *Law and Economics: New and Critical Perspectives*. New York: Peter Lang.
Cole, D. H. (1998). Instituting Environmental Protection. From Red to Green in Poland. London: Macmillan.
Commission of the European Communities (1995). *White Paper. Preparation of the Associated Countries of Central and Eastern Europe for Integration into the Internal Market of the Union*. Brussels.
Corbo, V., F. Coricelli, and J. Bossak (eds.) (1991). *Reforming Central and Eastern European Economies: Initial Results and Challenges*. Washington, DC: World Bank.
Cracow Academy of Economics, Grontmij Consulting Engineers, and National Foundation for Environmental Protection, Twente University (1995). *Development of Cost Methodologies and Evaluation of Cost-Effective Strategies for Achieving Harmonization with EC Environmental Standards*. Final Report. Kraków.
Cracow Academy of Economics, Grontmij Consulting Engineers, and National Foundation for Environmental Protection, Twente University (1996). *Development of Cost Methodologies and Evaluation of Cost-Effective Strategies for Achieving Harmonization with EC Environmental Standards*. Summary of the final report. Kraków.
Dąbrowski, B. (1994). "Śląski Czernobyl". *Wprost*, No. 36.
Darski, J., S. Szewczyk, and C. Maszczyk (1995). "Zagrożenia wodne dla kopalń czynnych na skutek likwidacji poszczególnych kopalń i ich części w obrębie Górnośląskiego Okręgu Przemysłowego". *Biuletyn PAWK*, No. 11.
Davies, N. (1986). *Heart of Europe. A Short History of Poland*. Oxford: Oxford University Press.
Davies, N. (1994). *Boże Igrzysko. Historia Polski. Tom 1*. Kraków.
Dawson, J. I. (1995). "Anti-Nuclear Activism in the USSR and its Successor States". *Environmental Politics*, Vol. 4, No. 3, Autumn, pp. 441-466.
Delorme, A. (1981). "Kryzys ekologiczny w Krakowie – próba krytycznej retrospekcji". *Aura*, No. 3.
DeBardeleben, J. (1985). *The Environment and Marxism-Leninism. The Soviet and East German Experience*. London: Westview Press.
De Esteban Lonso, A. and A. L. Lopez (1993). "Environmental Policy". In: A. A. Barbado (ed.), *Spain and EC Membership Evaluated*. London: Pinter.
Dittmer, L. (1983). "Comparative Communist Political Culture". *Studies in Comparative Communism*, No. 1 and 2, Spring/Summer, pp. 9-24.
Domański, J. (1981a). "W nurcie premian". *Aura*, No. 2.
Domański, J. (1981b). "Reforma gospodarcza bez ochrony środowiska". *Aura*, No. 3.
Domański, J. (1981c). "Zimno wszędzie, ciemno wszędzi?" *Aura*, No. 10.
Downing, P. B. (1984). *Environmental Economics and Policy*. Boston, MA: Little, Brown and Company.
Downing, P. B. and K. Hanf (eds.) (1983). *International Comparisons in*

Implementing Pollution Laws. The Hague: Kluwer Nijhoff.
Downs, A. (1972). "Up and Down with Ecology – the Issue-Attention Cycle". *The Public Interest*, Vol. 28, pp. 38-50.
Dryzek, J. S. (1987). *Rational Ecology: Environmental and Political Economy*. New York: Basil Blackwell.
Dudek, D., Z. Kulczyński, and T. Żylicz (1992). "Implementing Tradable Rights in Poland: A Case Study of Chorzów". Paper presented at the Third Annual Conference of the European Association of Environmental and Resource Economists, Cracow, 16-19 June.
Dudley, G. and J. Richardson (1996). "Why Does Policy Change over Time? Adversial Policy Communities, Alternative Policy Arenas, and British Trunk Roads Policy 1945-95". *Journal of European Public Policy*, March, No. 3: 1, pp. 63-83.
Dulewski, J. (1996). "Environmental legislation for the hard coal sector." Unpublished manuscript. Wyższy Urząd Górniczy, Katowice.
Dunn, W. N. (1981). *Public Policy Analysis. An Introduction*. Englewood Cliffs, NJ: Prentice Hall.
Eberg, J. (1997). *Waste Policy in Bavaria and the Netherlands. Policy Change and Learning*. Ph.D. Dissertation, University of Amsterdam.
Eckerberg, K. (1994). "Environmental Problems and Policy Options in the Baltic States: Learning from the West?" *Environmental Politics*, Vol. 3, No. 3, Autumn, pp. 445-478.
Ecofund (1992). *Ecofund Concept Paper*. Warsaw.
Economist (1981). "People Matter More Than Aluminium". January 10.
Edelman, M. (1977). *Political Language. Words that Succeed and Policies that Fail*. London: Academic Press.
Ekosol (1993). *Information Material on the Activities of Ekosol*. Kraków.
Ellman, M., E. Gaidar, and G. Kołodko (1993). *Economic Transition in Eastern Europe*. Cambridge, MA: Blackwell.
Enloe, C. (1975). *The Politics of Pollution in Comparative Perspective*. New York and London: Longman.
European Bank for Reconstruction and Development (EBRD) (1993). *Environmental Standards and Legislation in Western and Eastern Europe: towards Harmonization*. London.
Fagin, A. (1994). "Environment and Transition in the Czech Republic". *Environmental Politics*, Vol. 3, No. 3, pp. 479-494.
Familiec, J., K. Górka, and G. Mojżesz (1991). "Economic Losses due to Environmental Pollution in Poland and the Kraków Region". In: K. Górka (ed.), *Environmental and Economic Aspects of the Industrial Development in Poland*. Kraków: Cracow Academy of Economics, pp. 43-53.
Feshbach, M. and A. Friendly Jr (1992). *Ecocide in the USSR. Health and Nature under Siege*. New York: Basic Books.
Fiałkowski, T. (1996). "Nadzieja w gminach?" *Tygodnik Powszechny*, No. 46.
Ficek, M. and W. Ficek (1996). "Słone wody kopalniane a środowisko przyrodnicze". *Karbo-Energochemia-Ekologia*, No. 2.
Fiedor, B. (1995). "Ecological Aspects of Economic Relationships between Poland and the European Union". Unpublished paper. Wrocław.
Fischhoff, B. (1991). "Science and Politics in the Midst of Environmental Disaster". *Environment*, Vol. 33, No. 2, pp. 12-37.
Fisher, D. (1992). *Paradise Deferred: Environmental Policymaking in Central and

Eastern Europe. London: RIIA/Ecological Studies Institute.

Fitzmaurice, J. (1996). *Damming the Danube: Gabcikovo and Post-Communist Politics in Europe*. Boulder, CO: Westview Press.

Forowicz, K. (1992). "Kto truje – wie tylko GUS: stan przedzawałowy". *Rzeczpospolita*, 23/12.

Forowicz, K. (1996). "Odsalanie Wisły do 2002r". *Rzeczpospolita*, 18/4.

Forowicz, K. (1997). "Podziobie bocianów w Europie. Czarne i białe". *Rzeczpospolita*, 14-15/6.

French, H. F. (1988). "Industrial Wasteland". *World Watch*, Vol. 1, No. 6, November/December, pp. 21-30.

French, H. F. (1989). "The Greening of the Soviet Union". *World Watch*, Vol. 2, No. 3, May/June, pp. 21-29.

French, H. F. (1991). "Eastern Europe's Clean Break with the Past". *World Watch*, Vol. 4, No. 2, March/April, pp. 21-27.

Fyedorenko, N. P. (1976). "Przyroda w oczach ekonomisty". In: A. Kuklinski (ed.), *Człowiek i Środowisko*. Warszawa: Państwowe Wydawnictwo Ekonomiczne.

Gawlik, L. and I. Grzybek (1995). "Metodyka Oszacowania Emisji Metanu Związanego z Eksploatacją Węgla w Polsce". *Gospodarka Surowcami Mineralnymi*. Tom 11, Zeszyt 3.

Gelb, A. H. and C. W. Gray (1991). *The Transformation of Economies in Central and Eastern Europe: Issues, Progress, and Prospects*. The World Bank, Policy and Research Series. Washington, DC.

Gerrits, A. W. M. (1990). *The Failure of Authoritarian Change. Reform, Opposition, and Geopolitics in Poland in the 1980s*. Aldershot, UK: Dartmouth Publishing.

Gliński, P. (1993). "The Current State of Cooperation between NGOs and Government". In: P. Hardi, A. Juras and M. Tóth Nagy (eds.), *New Horizons? Possibilities for Cooperation between Environmental NGOs and Governments in Central and Eastern Europe*. Prague: WI&PO, pp.149-157.

Gliński, P. (1994). "Environmentalism among Polish Youth. A Maturing Social Movement?" *Communist and Post-Communist Studies*, No. 27. Vol. 2.

Gliński, P. (1995). *Polscy zieloni. Ruch społeczny w okresie przemian*. Warszawa.

Główny Urząd Statystyczny (GUS) (1989). *Ochrona Środowiska 1989*. Warszawa: ZWS.

Główny Urząd Statystyczny (GUS) (1991). *Ochrona Środowiska 1991*. Warszawa: ZWS.

Główny Urząd Statystyczny (GUS) (1995). *Ochrona Środowiska 1995*. Warszawa: ZWS.

Główny Urząd Statystyczny (GUS) (1997). *Ochrona Środowiska 1997*. Warszawa: ZWS.

Gofman, K. G. (1976). "Metodologiczne problemy optymalizacji procesów wykorzystywania przyrody w gospodarce socjalistycznej". In: A. Kukliński (ed.), *Człowiek i Środowisko*. Warszawa: Państwowe Wydawnictwo Ekonomiczne.

Górka, K. (1981). "Nakłady gospodarcze na ochronę środowiska". *Aura*, No. 9.

Górka, K. (1984). "Koszty ochrony środowiska w przemyśle". *Aura*, No. 3.

Górka, K. (1985). *Racjonalne wykorzystanie zasobów środowiska naturalnego i jego ochrona w procesie kształtowania rozwoju przemysłu*. Zeszyty Naukowe. Seria Specjalna: Monografie, No. 67, Akademia Ekonomiczna w Krakowie. Kraków.

Górka, K. (1991). "Changes in Environmental Policy in Poland". *The Shogaku Ronshu*, Vol. 59, No. 5, March, pp. 15-29.

Górka, K. (1994). "The Role of the State in Industrial Restructuring in Poland". Cracow Academy of Economics, Department of Industrial and Environmental Policy.

Górka, K. and B. Poskrobko (1987). *Ekonomia ochrony środowiska*. Warszawa: Państwowe Wydawnictwo Ekonomiczne.

Graaf, H. van de and R. Hoppe (1989). *Politiek en Belied*. Muiderberg: Coutinho.

Greenspan Bell, R. (1992a). "Industrial Privatization and the Environment in Poland". *Environmental Law Reporter*, No. 2, pp. 10092-10098.

Greenspan Bell, R. (1992b). "Environmental Law Drafting in Central and Eastern Europe". *Environmental Law Reporter*, Vol. 22, No. 9, pp. 10597-10606.

Grin, J. and R. Hoppe (1997). "Towards a Theory of the Policy Process: Problems, Promises and Prospects of the ACF". Paper presented at a workshop at the Department for Political Science, University of Amsterdam. Amsterdam, 4 February.

Haas, P., M. (1990). "Obtaining International Environmental Protection through Epistemic Consensus". *Millennium: Journal of International Studies*, Vol. 19, No. 3, pp. 347-363.

Haas, P. M., R. O. Keohane, and M. A. Levy (eds.) (1993). *Institutions for the Earth. Sources of Effective International Environmental Protection*. Cambridge, MA: MIT Press.

Hägerhäll, B. (ed.) (1988). *Vår gemensamma framtid*. Stocholm: Prisma/Tiden.

Hägerhäll, B. (1992). "Varför trivs storken i Baltikum?" Paper presented at Uppsala University, Sweden, 26 September.

Haggard, S. and S. B. Webb (1993). "What Do We Know about the Political Economy of Economic Policy Reform?" *The World Bank Research Observer*, Vol. 8, No. 2, July.

Havlicek, P. (1997). "The Czech Republic. First Steps Toward a Cleaner Future". *Environment*, Vol. 39, No. 3, pp. 16-41.

Hedlund, S. (1990). *Östeuropas kris*. Stockholm: Norstedts Förlag.

Heidenheimer, A. J., H. Heclo, and C. T. Adams (1990). *Comparative Public Policy: the Politics of Social Choice in America, Europe, and Japan*. New York: St. Martin's Press.

Held, D. (1987). *Models of Democracy*. Cambridge: Polity.

Hicks, B. (1996). *Environmental Politics in Poland. A Social Movement between Regime and Opposition*. New York: Columbia University Press.

Hisschemöller, M. (1993). *Demokratie van Problemen*. Amsterdam: VU Uitgiverij.

Hisschemöller, M. (1994). "The Democracy of Problems. Exploring the Relationship between Policy Content and Method of Political Decision". Paper submitted to the IPSA XVL World Congress, Berlin, 21-25 August, 1994.

Hisschemöller, M. and R. Hoppe (1996)."Coping with Intractable Controversies: The Case for Problem Structuring in Policy Design and Analysis". *Knowledge and Policy*, Vol. 8, No. 4, Winter, pp. 40-60.

Hjorth, R. (1992). *Building International Institutions for Environmental Protection. The Case of the Baltic Sea Environmental Cooperation*. Ph.D. Dissertation, Water and Environmental Studies, Linköping University, Sweden.

Hogwood, B. and L. Gunn (1984). *Policy Analysis for the Real World*. Oxford: Oxford University Press.

Hoogerwerf, A. (1990). "Reconstructing Policy Theory". *Evaluation and Program Planning*, Vol. 13, pp. 285-291.

Hoppe, R. and A. Peterse (1993). *Handling Frozen Fire: Political Culture and Risk*

Management. Boulder, CO: Westview Press.
Huber, G., S. Schönherr, and B. Thanner (1992). *Kazakhstan – an Economy in Transition*. München: Ifo Institute for Economic Research.
Hughes, G. (1990). "Are the Costs of Cleaning Up Eastern Europe Exaggerated? Economic Reform and the Environment". *CEPR Discussion Paper*, No. 482. November.
Hycnar, J. J. (1995). "Aspekty ekologiczne w procesach wydobywania, wzbogacenia i użytkowania węgla kamiennego". *Karbo. Energochemia. Ekologia*. Luty. No. 2.
ING BH Consultants, Poland and International Economic & Energy Consultants, UK (1994). *Financial and Economic Study of Environmental and Other Investment Projects for the Nadwiślańska Hard Coal Mines*. Final Report for PHARE Environmental Programme. Warszawa.
Institute for Sustainable Development (ISD) (1992). *Sustainable Development for Poland. Report by Polish Environmental NGOs, "Brazil 92"*. Warszawa.
Institute for Sustainable Development (ISD) (1998). *Comments to the National Integration Strategy*. Unpublished paper. Warszawa.
Instytut na rzecz Ekorozwoju (1993). *Ekologiczny okrągły stół po trzech latach*. Biuletyn No. 1. Warszawa.
Instytut Badań Klasy Robotniczej (1987). *Zagrożenie ekologiczne*. Warszawa.
International Union for Conservation of Nature and Natural Resources Conservation (IUCN) (1991). *The Environment in Eastern Europe*. Oxford.
Jacyna, I. (1987). *Czy zadepczemy przyrodę?* Warszawa: Nasza Księgarnia.
Jancar, B. (1987). *Environmental Management in the Soviet Union and Yugoslavia: Structure and Regulation in Federal Communist States*. Durham, NC: Duke University Press.
Jancar-Webster, B. (ed.) (1993). *Environmental Action in Eastern Europe: Responses to Crises*. New York and London: M.E. Sharp.
Jancar-Webster, B. (1995). "Environmental Degradation in Central Europe". In: J. DeBardeleben and J. Hannigan (eds.), *Environmental Security and Quality After Communism*. Oxford: Westview Press.
Jehlicka, P. and T. Kostelecky (1992). "The Development of the Czechoslovak Green Party Since the 1990 Elections". *Environmental Politics*, Vol. 1, No. 1, Spring, pp. 72-94.
Jendrośka, J. (1990). *State of Environmental Law: Poland*. Washington, DC: Center for International Environmental Law.
Jendrośka, J. (1992). "Environmental Law in Poland in a Transition Period: Recent Development of Legislation". *Tijdschrift voor Milieu & Recht*, October.
Jendrośka, J. (1996). "Drafting New Environmental Law in Poland: Radical Change or Merely Reform". In: G. Winter (ed.), *European Environmental Law*. Dartmouth: Aldershot.
Jendrośka, J. (1997). "Recent Development of Environmental Legislation in Poland in the Light of EU Directives". Proceedings of the International Conference on EU Directives and Environmental Law, Prague 13-14 May, pp. 104-108.
Jendrośka, J. and W. Radetzki (1991). "The Environmental Protection Act of 1980 – An Overview and Critical Assessment". In: Z. Bochniarz and R. Bolan (eds.), *Designing Institutions for Sustainable Development: A New Challenge for Poland*. Minneapolis (USA)-Białystok, Poland: Białystok Technical University, pp. 57-91.
Jenkins, T. (1991). "The Changing Face of Central and Eastern Europe and the Role of the Environmental Movement". *European Environment*, Vol. 1, Part 4,

August, pp. 1-4.
Juhasz, F. and A. Ragno (1993). "The Environment in Eastern Europe: From Red to Green?" *The OECD Observer*, No. 181, April/May.
Jurkiewicz, A. and J. Matiakowska (1984). "Śląsk przed zawałem". *Aura*, No. 1.
Kabala, S. J. (1988). *The State of Environmental Affairs in Poland: Regulation, Activism, and Public Concern in a Centrally Planned System*. Pittsburgh, PA.
Kamiński, B. (1994). "The Institutional Dimension of the Transition from Communism". *PPRG Discussion Papers*, No. 26. Warszawa: Department of Economics, Warsaw University.
Kapała, J. (1983). *Ocena jakości powietrza w województwie katowickim*. Wrocław: Zakład Narodowy im. Ossolińskich.
Kara, J. (1992). "Geopolitics and the Environment: The Case of Central Europe". *Environmental Politics*, Vol. 1, No. 2, Summer, pp. 186-195.
Karaczun, Z. (1993). *Policy of Air Protection in Poland. Part I and II*. Warszawa: Institute for Sustainable Development.
Karaczun, Z. (1994). *Policy of Air Protection in Poland. Part III*. Warszawa: Institute for Sustainable Development.
Karaczun, Z. and L. Indeka (1996). *Ochrona Środowiska*. Warszawa: Aries.
Karbownik, A. (1994). "Problem ochrony środowiska w górnictwie węgla kamiennego w Polsce". *Wiadomści Górnicze*, No. 9.
Karbownik, A. (1995). "Ochrona środowiska w górnictwie węgla kamiennego w latach 1990-1994". *Wiadomości Górnicze*, No. 6.
Karczewski, H. (1984). "Proces trwa – jak zapadnie wyrok". *Aura*, No. 2.
Karkoszka, K. (1996). "Sól z ziemi czarnej". *Trybuna Górnicza*, No. 25.
Karlsson, G. (1996). "På 7 sekunder gick reaktorn över styr". *Dagens Nyheter*, 19 April.
Keegan, W. (1993). *The Spectre of Capitalism. The Future of the World Economy after the Fall of Communism*. London: Vintage.
Keren, M. and G. Ofer (eds.) (1993). *Economic Reform in the Former Communist Bloc*. Boulder, CO: Westview Press.
Kingdon, J. W. (1984). *Agendas, Alternatives, and Public Policies*. New York: Harper Collins.
Klaus, V. (1992). Dismantling Socialism: A Preliminary Report (A Road to Market Economy II). Praha: Top Agency.
Kleijn, A. (1968). *Beleid en Wetenschap*. Alphen aan de Rijn.
Klok, P-J (1991). *Een instrumententheorie voor milieubelied*. Ph.D. Dissertation, Twente University.
Kochel, Z. and L. Pinko (1996). "Dokonania energetyki zawodowej w inwestycjach proekologicznych". *Biuletyn Energetyczny*, Izba Gospodarcza Energetyki i Ochrony Środowiska. No. 2.
Kołodko, G. W. and D. M. Nuti (1997). *The Polish Alternative. Old Myths, Hard Facts, and New Strategies in the Successful Transformation of the Polish Economy. Research for Action 33*. Helsinki: The United Nations University, World Institute for Development of Economics Research.
Komisja do Spraw Ekorozwoju (1996). *Propozycja tematów pracy Komisji do Spraw Ekorozwoju w roku 1996*. Memo. Warszawa.
Komisja do Spraw Ocen Oddziaływania na Środowisko (1996). Information card on the planned desalination plant in Oświęcim. Warszawa.
Komisja Planowania przy Radzie Ministrów (1988). *Ocena przebiegu i wyników wdrażania reformy gospodarczej w 1987 roku*. Warszawa.

Komisja Planowania Przy Radzie Ministrów (1986). *Założenia planu przestrzennego zagospodarowania kraju do 1995 roku*. Warszawa.

Kornai, J. (1992). *The Socialist System. The Political Economy of Communism*. Oxford: Clarendon Press.

Köves, A. (1992). *Central and East European Economies in Transition: The International Dimension*. Boulder, CO: Westview Press.

Kozłowski, S. (1981). "Jak zarządzać środowiskiem?" *Aura*, No. 9.

Kozłowski, S. (ed.) (1986). *Expertyzy – opinie o wpływie inwestycji na środowisko*. Polska Akademia Nauk. Komitet Przestrzennego Zagospodarowania Kraju. Biuletyn Zeszyt 129. Warszawa: Państwowe Wydawnictwo Naukowe.

Kozłowski, S. (1992). "Opening Address", High-Level Meeting on Co-operation and Sustainable Development in the Chemical Industry. Warszawa, March.

Kozłowski, S. (1994). *Droga do Ekorozwoju*. Warszawa.

Kozłowski, S. (1995). "Przyrodnicze kryteria gospodarki przestrennej w świetle ustawy o zagospodarowaniu przestrzennym z 1994 roku". Unpublished material. Warszawa.

Krajowa Komisja Ochrony Środowiska Człowieka NSZZ Solidarność (1981). "O zdrowie narodu, o szacunek dla piekna i bogactwa Polski". *Aura*, No. 10.

Kramer, J. M. (1995). "Energy and the Environment in Eastern Europe". In: J. DeBardeleben and J. Hannigan (eds.), *Environmental Security and Quality after Communism*. Oxford: Westview Press.

Kroner, J. (1992). "Trybunał Konstitycyjny bez togi". *Rzeczpospolita*, 29/7.

Kukliński, A. (1976). *Człowiek i Środowisko*. Warszawa: Państwowe Wydawnictwo Ekonomiczne.

Lane, J-E (1995). *The Public Sector. Concepts, Models, and Approaches*. London: Sage.

Laurson, P., A. Melzer, and T. Zylicz (1995). *A Strategy to Enhance Partnerships in Project Financing for Environmental Investments in Central and Eastern Europe*. Report for the European Bank for Reconstruction and Development. London.

Lavigne, M. (1995). *The Economics of Transition. From Socialist Economy to Market Economy*. London: Macmillan.

Lenssen, N. and C. Flavin (1996). "Meltdown". *World Watch*, Vol. 9, No. 3, May/June, pp. 23-31.

Lester, J. P. (ed.) (1989). *Environmental Politics and Policy. Theories and Evidence*. Durhan, NC and London: Duke University Press.

Lewiński, A. (1981). "Nie zabijaj środowiska". *Aura*, No. 10.

Liefferink, J. D. (1995). *Environmental Policy on the Way to Brussels. The Issue of Acidification between the Netherlands and the European Community*. Ph.D. Dissertation. Den Haag.

Lindblom, C. E. (1965). *The Intelligence of Democracy. Decision-Making Through Mutual Adjustment*. New York: Free Press.

Lindblom, C. E. (1980). *The Policy-Making Process*. Englewood Cliffs, NJ: Prentice Hall.

Lipszyc, J. B. (1980). "Nie jesteśmy na państowym garnuszku". *Przegląd Techniczny*, No. 45.

Loeber, A. and J. Grin (1996). "From Green Waters to Green Detergents: Processes of Learning between Policy Actors and Target Groups in Eutrophication Policy in the Netherlands, 1970-1987". Paper presented at the 1996 WPSA Conference.

Lubbe, A. (1991). "Transforming Poland's Industry". *PPRG Discussion Papers*, No.

13. Department of Economics, Warsaw University.
Lundqvist, L. (1980). *The Hare and the Tortoise: Clean Air Policies in the United States and Sweden*. Ann Arbor, MI: University of Michigan Press.
Maciejewski, S. (1980). "Do sejmu Polskiej Rzeczypospolitej". *Aura*, No. 12.
Maciejewski, E. (1994). "Ekologiczna gospodarka rynkową szansa rozwoju kraju". *Środowisko*, No. 11.
Malawski, M. (1987). "Od systemu dyrektyw do rynku uprawnień". *Aura*, No. 8, pp. 6-7.
Manser, R. (1994). *Failed Transitions*. New York: New Press.
Marples, D. R. (1986). "Chernobyl and Ukraine". *Problems of Communism*, November-December, pp. 17-27.
Massam, B. H. and R. Earl-Goulet (1997). "Environmental Nongovernmental Organizations in Central and Eastern Europe. Contributions to Civil Society". *International Environmental Affairs*, Vol. 9, No. 2, Spring, pp. 127-147.
Matuszewska, E. and J. Spyrka (1995). "Poland". In: *Status of National Environmental Action Programmes in Central and Eastern Europe*. Budapest: The Regional Environmental Centre.
Mazmanian, D. A. and P. A. Sabatier (1989). *Implementation and Public Policy*. New York: University Press of America.
McCormick, J. (1992: 1989). *The Global Environment Movement*. London: Belhaven Press.
Mielczarek, A. (1983). "O elektownej bez emocji". *Aura*, No. 8.
Miljödepartementet (1997). *Miljön i ett utvidgat EU. Betänkande av kommitten om EUs utvidgning – analys av miljökonsekvenserna*. SOU, No 149. Stockholm.
Millard, F. (1994). *The Anatomy of the New Poland. Post-Communist Politics in its First Phase*. Aldershot, UK: Edward Elgar.
Millard, F. (1998). "Environmental Policy in Poland". *Environmental Politics*, Vol. 7, No. 1, pp. 145-161.
Ministerstwo Gospodarki Terenowej i Ochrony Środowiska (MGTiOŚ) (1973). *Kompleksowy program ochrony i kszałtowania środowiska w Polsce do roku 1990*. Warszawa.
Ministerstwo Gospodarki Terenowej i Ochrony Środowiska (MGTiOŚ) (1975). *Założenia programu ochrony środowiska w Polsce do roku 1990*. Warszawa.
Ministerstwo Ochrony Środowiska i Zasobów Naturalnych (MOŚiZN) (1988a). *Narodowy program ochrony środowiska przyrodniczego do roku 2010. Materiały do konsultacji społecznej*. Warszawa: Wydawnictwo Geologiczne.
Ministerstwo Ochrony Środowiska i Zasobów Naturalnych (MOŚiZN) (1988b). *Narodowy program ochrony środowiska przyrodniczego do roku 2010. Projekt*. Warszawa: Wydawnictwo Geologiczne.
Ministerstwo Ochrony Środowiska i Zasobów Naturalnych (MOŚiZN) (1988c). Unpublished material on the structure of the Ministry for Environmental Protection and Natural Resources. Warszawa.
Ministerstwo Ochrony Środowiska, Zasobów Naturalnych i Leśnictwa (MOŚZNiL) (1997). *Przegląd postępu poczynionego od czasu UNCED*. Report for the UN Commission for Sustainable Development. June. Warszawa.
Ministerstwo Przemysłu i Handlu (1996). *Górnictwo węgla kamiennego. Polityka państwa na lata 1996-2000. Program dostosowania górnictwa węgla kamiennego do warunków gospodarki rynkowej i międzynarodowej konkurencyjności*. Katowice: Wydawnictwo Górnicze.
Ministry of Environmental Protection (1998). *Baltic 21. Transport Sector*. Baltic 21

Series, No. 8/98. Stockholm.
Ministry of Environmental Protection, Natural Resources, and Forestry (MEPNRaF) (1991). *National Environmental Policy of Poland*. Warszawa.
Ministry of Environmental Protection, Natural Resources, and Forestry (MEPNRaF) (1995). *National Environmental Action Programme*. Warszawa.
Ministry of Environmental Protection, Natural Resources, and Forestry (MEPNRaF) (1996). Unpublished statistics on saline water discharges. Warszawa.
Ministry of Environmental Protection, Natural Resources, and Forestry (MEPNRaF) (1997). *The Status of Approximation of the Polish Environmental Law to the Environmental Law of the European Union*. Unpublished memo. Warszawa.
Ministry of Environmental Protection, Natural Resources, and Forestry (MEPNRaF). *Environmental Protection;* (1998a). Unpublished memo on adjustment of Polish environmental law to EU legislation. Warszawa.
Ministry of Environmental Protection, Natural Resources, and Forestry (MEPNRaF). *Environmental Protection* (1998b). Unpublished memo on adjustment of Polish environmental law to EU legislation. Annex. Warszawa.
Mintrom, M. and S. Vergari (1996). "Advocacy Coalitions, Policy Entrepreneurs, and Policy Change". *Policy Studies Journal*, Vol. 24, No. 3, pp. 420-434.
Mistewicz, E. (1997). "Zielona alternatywa". *Wprost*, No. 38.
Mnatsakian, R. A. (1992). *Environmental Legacy of the Former Soviet Republics*. Edinburgh: Centre for Human Ecology, University of Edinburgh.
Moltke, K. von (1993). *Environmental Standards and Legislation in Western and Eastern Europe: Towards Harmonization*. Institute for European Environmental Policy, Arnhem.
Moltke, K. von (1995). *Regulatory Trends in OECD Countries: Issues for Trade and Harmonization*. China Council for International Cooperation on Environment and Development.
Motyka, I. and A. Magdziorz (1994). "Kierunki i postęp w odsalaniu wód". *Wiadomości Górnicze*, No. 9.
Myrdal, G. (1944, 1966). *An American Dilemma. The Negro Problem and Modern Democracy*. New York: Harper & Row.
Nadwiślańska Spółka Węglowa SA (NSW) (1994). Information material on NSW. Katowice.
Nadwiślańska Spółka Węglowa SA (NSW). (1996). *Informacja. Program zagospodarowania zasolonych wód dołowych z kopalń Nadwiślańskiej Spółki Węglowej* SA. Tychy. Katowice.
National Foundation for Environmental Protection (NFEP) (1997). *Agenda 21 in Poland. Progress Report 1992-1996*. Warszawa.
National Fund for Environmental Protection and Water Management (NFEPaWM) (1996). *The National Fund for Environmental Protection and Water Management. Assistance for Eco-Development*. Warszawa.
National Fund for Environmental Protection and Water Management (NFEPaWM) (1997). *Eight Years of Operation. A Report Summary*. Warszawa.
Nelson, T. (1996). "Russia's Population Sink". *World Watch*, Vol. 9, No. 1, January/February, pp. 22-23.
Nilsson, J. (1987). "Putting it in Figures". In: *The Limits to Nature's Tolerance*. Report from the NGO Conference on Acid Rain, Lida, Stockholm, Sweden, 6-7 September.
North, D. C. (1996). *Institutions, Institutional Change, and Economic Performance*. Cambridge: Cambridge University Press.

Nove, A. (1980, 1987). *The Soviet Economic System*. London: George Allen & Unwin.

Novy, M. (1996). "International Actions to Combat Pollution in the Black Triangle and Katowice Regions". Paper presented at Second Rhine/Katowice Policy Comparison Workshop, IIASA, Laxenburg, Austria, 24-25 April.

Nowicki, M. (1993). *Environment in Poland. Issues and Solutions*. Dordrecht: Kluwer.

Ochocki, L. (1985). *Ochrona środowiska. Problemy, refleksje, dokumenty*. Warszawa: Wydawnictwo Stronnictwa Demokratycznego Epoka.

Organization for Economic Cooperation and Development (OECD) (1993). *Environmental Performance Reviews: Germany*. Paris.

Organization for Economic Cooperation and Development (OECD) (1994a). *Managing the Environment. The Role of Economic Instruments*. Paris.

Organization for Economic Development and Cooperation (OECD) (1994b). *Reducing Environmental Pollution: Looking Back, Thinking Ahead*. Paris.

Organization for Economic Development and Cooperation (OECD) (1995a). *Environmental Performance Reviews: Poland*. Paris.

Organization for Economic Cooperation and Development (OECD) (1995b). *Environmental Funds in Economies in Transition*. Paris.

Organization for Economic Cooperation and Development (OECD) (1997). *Environmental Performance Reviews: Belarus*. Paris.

O'Riordan, T. (1995). "Frameworks for Choice. Core Beliefs and the Environment." *Environment*, Vol. 37, No. 8.

Panek, K. (1992). "Zielone prawo wspólnoty". *Przegląd Techniczny*, No. 29.

Parker, M. (1994). *The Politics of Coal's Decline. The Industry in Western Europe*. London: Earthscan Publications.

Pavlinek, P. (1995). *Transition and the Environment in the Czech Republic: Democratization, Economic Restructuring, and Environmental Management in the Most District after the Collapse of State Socialism*. Ph.D. Dissertation, University of Kentucky, Lexington.

Pazdan, R. (1996). "Tradable Emission Permit System". In: Ministry of Environmental Protection, Natural Resources, and Forestry (ed.), *Proceedings from Marketable Permit Workshop*. Jadwisin, Poland, 18-19 June.

Peszko, G. (1990). "Green Federation – A Political Movement for Radical Ecology". Unpublished manuscript. Kraków.

Peterson, D. J. (1993). *Troubled Lands. The Legacy of Soviet Environment Destruction*. Boulder, CO.: Westview Press.

Pickel, A. (1993). "Authoritarianism or Democracy? Marketization as a Political Problem". *Policy Sciences*, No. 26, pp. 139-163.

Piontek, F. (1988). "Ocena wartości ekonomicznej strat spowodowanych brakiem skutecznej ochrony środowiska." In: *Problemy rozwoju społeczno-gospodarczego z poszanowaniem dóbr przyrody*. Wrocław: Polska Akademia Nauk.

Polski Klub Ekologiczny (PKE) (1986). *Ekorozwój szansą przetrwania cywilizacji*. Kraków: Wydawnictwo AGH.

Polski Klub Ekologiczny (PKE) (1987). *Dokument Polskiego Klubu Ekologicznego. Deklaracja ideowa, tezy programowe, uchwały*. Zarząd Główny PKE, Kraków.

Polityka (1980). "Protokół porozumienia zawartego przez komisję rządową i międzyzakładowy komitet strajkowy w dn. 31 sierpnia 1980 r. w stoczni Gdańskiej". No. 26, 6/9.

Prandecka, B. (1980). *Polityka ochrony środowiska w krajach socjalistycznych*. Warszawa: Państwowe Wydawnictwo Ekonomiczne.

Premfors, R. (1989). *Policyanalys*. Lund, Sweden: Studentlitteratur.

Pressman, J. and A. Wildawsky (1973, 1984). *Implementation*. Berkeley, CA: University of California Press.

Pryde, P. R. (1991). *Environmental Management in the Soviet Union*. Cambridge: Cambridge University Press.

Pryde, P. R. (ed.) (1995). *Environmental Resources and Constraints in the Former Soviet Republics*. Boulder, CO: Westview Press.

Przyroda Polska (1981). "Raport Ligi Ochrony Przyrody o stanie środowiska przyrodniczego w Polsce i zagrożeniu zdrowia ludzkiego". No. 5/6.

Rada Ekologiczna przy Prezydencie Rzeczypospolitej Polskiej (1993). *Biuletyn*, No. 1. Warszawa.

Radecki, W. (1979). "Ochrona środowiska w ustawach zasadniczych państw socjalisty-cznych. Dobro konstytucyjnie chronione". *Aura*, No. 6.

Radecki, W. (1983). "Fundusze na ochronę.środowiska". *Aura*, No. 4.

Radecki, W. (1992). *Ustawa o Państwowej Inspekcji Ochrony Środowiska*. Wrocław.

Radetzki, M. (1995). *Polish Coal and European Energy Market Integration*. Aldershot, UK: Avebury.

Radetzki, M. (1997). *Polish Hard Coal – the Road to Economic Health*. Stockholm: SNS Energy Institute of Sweden.

Reeves, A. (1995). "Towards Practical Environmental Provision in Hungary". *European Environment*, Vol. 5, pp. 70-75.

Regional Environmental Center for Central and Eastern Europe (REC) (1995). *Status of National Environmental Action Programs in Central and Eastern Europe*. Budapest.

Regional Environmental Center for Central and Eastern Europe (REC) (1996). *Approximation of European Union Environmental Legislation*. Budapest.

Regionalny Zarząd Gospodarki Wodnej w Krakowie (1996). *Ciągly pomiar zasolenia wody w rzece Wiśle na stopniu wodnym Kościuszko*. No. 5.

Review of European Community & International Environmental Law (RECIEL) (1994). "EBRD's Municipal and Environmental Development Initiative". Vol. 3, No, 4, pp. 289-291.

Ribeiro, J. (1996). *Desalination Technology. Survey and Prospects*. Institute for Prospective Technological Studies, Seville.

Richardson, J. (ed.) (1982). *Policy Styles in Western Europe*. London: George Allen & Unwin.

Richter, S. (ed.) (1992). *The Transition from Command to Market Economies in East-Central Europe*. The Vienna Institute for Comparative Economic Studies Yearbook IV. Boulder, CO: Westview Press.

Rogóż, M. (1994). "Słone wody kopalniane w Górnośląskim Zagłębiu Węglowym i możliwość ich wtłaczania do górotworu". *Wiadomości Górnicze*, No. 9.

Rose, R. (1993). *Lesson-Drawing in Public Policy*. Chatham, NJ: Chatham House.

Ross, L. (1988). *Environmental Policy in China*. Indianapolis, IN: Indiana University Press.

Rothstein, B. (1994). *Vad bör staten göra? Om välfärdsstatens moraliska och politiska logik*. Stockholm: SNS Förlag.

Równy, K. (1996). "Ekologiczne warunki członkowstwa Unii Europeiskiej i zbliżanie się do nich w Polsce". *Prawo i Środowisko*. No. 3, pp. 13-23.

Rzeczpospolita (1998a). "Górnictwo: mniej osob do zwolnienia". 23-24/5.

Rzeczpospolita (1998b). "Najtrudniej w przemyśle ciężkim". 25-26/4.
Rzeczpospolita (1998c). "Stopa bezrobocia spadła do 10.2 proc.". 23-24/5.
Sabatier, P. A (1998). "The Advocacy Coalition Framework: Revisions and Relevance for Europe". *Journal of European Public Policy*, Vol. 5, No. 1, March, pp. 98-130.
Sabatier, P. A. and H. C. Jenkins-Smith (eds.) (1993). *Policy Change and Learning. An Advocacy Coalition Approach*. Boulder, CO: Westview Press.
Sabatier, P. A. and H. C. Jenkins-Smith (1999). "The Advocacy Coalition Framework: An Assessment". In: Sabatier, P. (ed.). *Theories of the Policy Process*. Boulder, CO: Westview Press.
Sachs, J. (1994). *Poland's Jump to the Market Economy*. Cambridge, MA: MIT Press.
Salay, J. (1991). *Östeuropas miljö. Problem och framtidsutsikter*. Falköping, Sweden: Naturia.
Salay, J. (1996). *Energy Use and Air Pollution in a Transitional Industrialized Economy. The Case of Poland*. Lund University, Sweden.
Sätre-Åhlander, A-M (1988). *Miljöproblemen i Sovjetunionen*. Stockholm: Utrikespolitiska Institutet.
Sätre-Åhlander, A-M (1993). *The Effect of the Soviet Shortage Economy on the Environment and the Use of Natural Resources*. Ph.D. Dissertation. Department of Economics, Stockholm University.
Schattschneider, E. E. (1960). *The Semisovereign People. A Realist's View of Democracy in America*. New York: Holt, Rinehart & Winston.
Schmitter, P. C. (1979). "Still the Century of Corporatism?" In: P. C. Schmitter and G. Lehmbruch (eds.), *Trends towards Corporatist Intermediation*. London: Sage.
Schreiber, H. (1990). "The Continent's Conundrum". *The Environmental Forum*. September-October.
Semeniene, D. and T. Żylicz (1997). "Implementing Environmental Policies - the Role of Economic Instruments". In: T. Żylicz (ed.), *Ecological Economics: Markets, Prices and Budgets in a Sustainable Society*. The Baltic University Programme, Session 8. Uppsala: Uppsala Publishing House.
Serafin, R. (1982). "The Greening of Poland". *Ecologist*, Vol. 12, No. 4.
Sikora, A. (1988). *Ochrona Bałtyku i jego zasobów*. Warszawa: Ludowa Spółdzielnia Wydawnicza.
Sikora, J., K. Szyndler, and R. Ludlum (1994). "Zakład odsalania przy kopalni Dębieńsko. Bezodpadowa utylizacja zasolonych wód kopalnianych". *Wiadomści Górnicze*, No. 9.
Simpson, P. (1996). "Katowice". *Business Central and Eastern Europe*. Vol. 4, No. 37.
Śleszyński, J. (1996a). "Economic Instruments in Environmental Policy: A Profile of Poland". *European Environment*, Vol. 6, pp. 126-134.
Śleszyński, J. (1996b). "The Polish Environmental Policy. Two Case Studies on Implementation of Selected Economic Instruments". Unpublished paper prepared for the Harvard Institute for International Development.
Sommer, J. (1993). "Poland". In: R. and M. Boes (eds.), *International Encyclopaedia of Laws*. Deventer: Kluwer Law and Taxation Publishers.
Sörlin, S. (1997). "The Environmental Dilemma - a History of Scientists and Social Constructions". In: S. Sörlin (ed.), *The Road towards Sustainability. A Historical Perspective*. Session 1. The Baltic University Programme. Uppsala: Uppsala Publishing House.

State Inspectorate for Environmental Protection (1996). *The Concept of Compliance Programmes*. Warsaw.

Stavins, R. N. and T. Żylicz (1995). "Environmental Policy in a Transition Economy: Designing Tradeable Permits for Poland". *Harvard Institute for International Development. Environmental Discussion Paper*. No. 9.

Steenge, A. E. (1991). "Environmental Economics in the Formerly Centrally Planned Economies in Eastern Europe. Some Observations". Paper presented at the Annual Meeting of the European Association of Environmental and Resource Economists, Stockholm, 10-14 June.

Stetson, M. (1990). "Auf Widersehen East German Nukes?" *World Watch*, Vol. 3, No. 5, September/October, p. 35.

Stoklasa, J. (1980). "Rozwój społeczno-gospodarczy a ochrona środowiska". In: B. Prandecka (ed.), *Polityka ochrony środowiska w krajach socjalistycznych*. Warszawa: Państwowe Wydawnictwo Ekonomiczne.

Swedish-Polish Association for Environmental Protection (SPM) (1989). *Proceedings from SPM Seminar on Swedish-Polish Environmental Cooperation*. 23-24 February 1988. Uppsala.

Swolkień, O. (1995). *Nowy ustrój - te same wartości*. Kraków: Biblioteka Zielonych Brygad.

Szamałek, K. (1996). "Marketable Pollution Permits". In: Ministry of Environmental Protection, Natural Resources, and Forestry (ed.), *Proceedings from Marketable Permit Workshop*, Jadwisin, Poland, 18-19 June.

Szwed, D. (1996). "Ekologiczna reforma podatkowa". *Środowisko*, No. 20.

Timberlake, L. (1981). "Poland - the Most Polluted Country in the World?" *New Scientist*, 22 October.

Tkaczyński, J. K. (1997). "Von der Polnischen Wirtschaft zum Polnischen Tiger Europas?" *Osteuropa-Wirtschaft*, Vol. 3, No. 3.

Tobera, P. (1981). "Świadomość ekologiczna i jej uwarunkowania". *Aura*, No. 7.

Tobera, P. (1988). *Kryzys środowiska - kryzys społeczeństwa*. Warszawa: Ludowa Spółdzielnia Wydawnicza.

Tolba, M. K., O. A. El-Kholy, E. El-Hinnawi, M. W. Holdgate, D. F. McMichael, and R. E. Munn (eds.) (1992). *The World Environment 1972-1992. Two Decades of Challenge*. London: Chapman & Hall.

Trybuna Górnicza (1996). "Decyzja KERM: mniej za słone wody". No. 48.

Turnbull, M. (1991). *Soviet Environmental Policies and Practices: The Most Critical Investment*. Aldershot and Brookfield, VT: Dartmouth Publishers.

UNCED/Estonia (1992). *National Report of Estonia to the United Nations Conference on Environment and Development*. Ministry of Environment, Tallin.

United Nations Department of Public Information (1993). *Earth Summit. Agenda 21. The United Nations Programme of Action from Rio*. New York: United Nations Reproduction Section.

Vedung, E. (1995). "Policy Instruments: Typologies and Theories". Unpublished manuscript, Uppsala University, Sweden.

Vogel, D. (1986). *National Styles of Regulation. Environmental Policy in Great Britain and the United States*. Ithaca, NY and London: Cornell University Press.

Wajda, S. (1993). "Improvement of Compliance with Environmental Legislation: The Case of Poland". United Nations Economic Commission for Europe, Workshop on Legal and Regulatory Framework for Environmental Management, Sofia 30 June to 2 July 1993, pp. 3-28.

Wajda, S. (1996a). "Approximation of Polish Environmental Law to European Union

Legislation". Unpublished manuscript. Ministry of Environmental Protection, Natural Resources and Forestry. Memo. Warszawa.

Wajda, S. (1996b). "Dostosowanie polskiego prawa do ustawodawstwa Unii Europejskiej". *Ekopartner*, No. 2.

Wajda, S. (1996c). "Przełom w prawie ochrony środowiska". *Prawo i Środowisko*, No. 3.

Wajda, S. (1996d). "Unia Europejska". *Ekopartner*, No. 3.

Waller, M. and F. Millard (1992). "Environmental Politics in Eastern Europe". *Environmental Politics*, Vol. 1, No. 2, Summer, pp. 159-185.

Weale, A. (1992). *The New Politics of Pollution*. Manchester: Manchester University Press.

Węcławowicz, G. (1996). *Contemporary Poland. Space and Society*. London: UCL Press.

Węgrzynowska, I. (1993). "Kierunki działań w zakresie utylizacji i zagospodarowania wód słonych z kopalń węgla kamiennego". In: Instytut Ekologii Terenów Uprzemy-słowionych. Program działań ekologicznych zmierzających do poprawy warunków życia mieszkańców Górnego Śląska. Katowice.

Weissenburger, U. (1995). "Umweltprobleme und Umweltschutz in Weissrussland". *Osteuropa-Wirtschaft*, Vol. 40, No. 1, pp. 13-23.

Welfens, M. J. (1993). *Umweltprobleme und Umweltpolitik in Mittel- und Osteuropa. Ökonomie, Ökologie und Systemwandel*. Heidelberg: Physica-Verlag.

Welford, R. (1991). "The Environmental Impact of German Reunification". *European Environment*, Vol. 1, Part 2, April, pp. 8-11.

Wijetilleke, L. and S. A. R. Karunaratne (1995). *Air Quality Management. Considerations for Developing Countries*. World Bank Technical Paper, No. 278. Washington, DC.

Winsemius, P. (1989). *Guests in Our Own Home*. Alphen aan den Rijn: Samson.

Wizelius, T. (1992). "Prognoserna från Tjernobyl som IAEA försökte dölja: en halv miljon döda inom 50 år". *Miljömagasinet Alternativet*, No. 51.

Wójcik, P. (1985). *Zagrożenie Ekologiczne*. Akademia Nauk Społecznych, Instytut Badań Klasy Robotniczej. Warszawa.

Wojewoda Katowicki (1985). *Projekt zadań wieloletniego programu ochrony i kształtowania środowiska w województwie katowickim na lata 1986-1990, 1991-1995*. Katowice.

World Bank (1989a). *Poland's Economic and Industrial Structure*. Background Paper. Washington, DC.

World Bank (1989b). *Sources of Pollution in Poland. Coal*. Background Paper. Washington, DC.

World Bank (1989c). *Pollution Levels - the Spatial Dimension*. Background Paper. Washington, DC.

World Bank (1989d). *The Costs of Environmental Degradation in Poland*. Background Paper. Washington, DC.

World Bank (1989e). *Participants in Environmental Protection*. Background Paper. Washington, DC.

World Bank (1989f). *Administration of Environmental Protection*. Background Paper. Washington, DC.

World Bank (1989g). *Sectoral Issues - Some Examples*. Background Paper. Washington, DC.

World Bank (1992a). *Czech and Slovak Federal Republic Joint Environmental Study*. Washington, DC.

World Bank (1992b). *Poland. Environmental Strategy*. Washington, DC.

World Bank (1992c). *Russian Economic Reform*. Washington, DC.

World Bank (1992d). *World Development Report 1992: Development and Environment*. Oxford.

World Bank (1992e). *The World Bank and the Environment*. Fiscal 1992. Washington, DC.

World Bank (1993a). *Estonia - the Transition to a Market Economy*. Washington, DC.

World Bank (1993b). *Lithuania - the Transition to a Market Economy*. Washington, DC.

World Bank (1993c). *Poland. Energy Sector Restructuring Programme. Volume 2. The Hard Coal Sector*. Washington, DC.

World Bank (1994). *Environmental Action Programme for Central and Eastern Europe*. Abridged version of the Document endorsed by the Ministerial Conference. Lucerne, Switzerland, 28-30 April 1993.

World Commission on Environment and Development (WCED) (1987). *Our Common Future*. Oxford: Oxford University Press.

World Resources Institute (1992). *World Resources 1992-93*. Oxford and New York: Oxford University Press.

World Wide Fund International (1993). Special issue on Central and Eastern Europe. March.

Wuester, H. (1996). "The Role of VOC Abatement Cost Estimates in the Preparations of a New Protocol under the Convention of Long-Range Transboundary Air Pollution". Secretariat of the United Nations Economic Commission for Europe. Geneva.

Yarnal, B. (1995). "Bulgaria at Crossroads. Environmental Impacts of Socioeconomic Change". *Environment*, Vol. 37, No. 10, pp. 6-29.

Yin, R. K. (1994). *Case Study Research. Design and Methods*. Second Edition. London: Sage.

Zarafescu, C. (1992). "The State of the Environment in Romania". *European Environment*, Vol. 2, Part 2, April, pp. 16-17.

Zaręba, J. (1983). "Z motyką na słońce". *Aura*, No. 2.

Zaręba, J. (1985). "Ustawa po pięciu latach". *Aura*, No. 10.

Ziegler, C. E. (1987). *Environmental Policy in the USSR*. London: Pinter.

Zielona Alternatywa (1991), No. 1.

Zielone Brygady (Green Brigades), all issues between 1989 and 1995.

Zimny, H. (1988). *Czym naprawdę oddychamy*. Warszawa: Krajowa Agencja Wydawnicza.

Żylicz, T. (1989). *Ekonomia wobec problemów środowiska przyrodniczego*. Warszawa: Państwowe Wydawnictwo Naukowe.

Żylicz, T. (1991). "The Greening of the Post-Communist Europe?" Paper presented at the Second Annual Conference of the European Association of Environmental and Resource Economists. Stockholm, 11-14 June.

Żylicz, T. (1992a). "Environmental Policy Reform in Poland". In: T. Sterner (ed.), *Economic Policies for Sustainable Development*. Dordrecht: Kluwer.

Żylicz, T. (1992b). "Debt-for-Environmental Swaps. The Institutional Dimension". *Beijer Discussion Papers*, No. 18. Stockholm: The Beijer Institute, The Royal Swedish Academy of Sciences.

Żylicz, T. (1992c). "It's Time for Economics". Unpublished paper, July. Warszawa.

Żylicz, T. (1992d). "Applied Environmental and Resource Economics in Poland since

1989". Unpublished paper. Warszawa.

Żylicz, T. (1995a). "Goals, Principles, and Constraints in Environmental Policy". In: H. Folmer, H. Landis Gabel, and H. Opschoor (eds.), *Principles of Environmental and Resource Economics. A Guide for Students and Decision-Makers*. Aldershot, UK: Edward Elgar.

Żylicz, T. (1995b). "Can There be an Ecological Tax?" *Economic Discussion Papers*, No. 17. Faculty of Economics, Warsaw University.

Żylicz, T. (1996). "Introduction". In: Ministry of Environmental Protection, Natural Resources and Forestry (ed.), *Proceedings from Marketable Permit Workshop*. Jadwisin, Poland, 18-19 June.

Żylicz, T. (1997). "Environmental Policy in Economies in Transition". In: *International Yearbook of Environmental and Resource Economics 1998/1999*, pp. 255-256.

Żylicz, T. (1998). "Obstacles to Implementing Tradable Pollution Permits in Poland". Unpublished paper, Warsaw Ecological Economics Center.

Interviews

Robert Alberski, Department for Political Science, Wrocław University. November 1995 and December 1996.

Hanna Baradziej, Department of Water Management, Ministry of Environmental Protection, Natural Resources, and Forestry. July 1996.

Wojciech Beblo, Chairman of the Katowice Branch of the Polish Ecological Club until 1991. Director for the Department for Ecology, Katowice Voivodship 1991-95. Since 1995: Commercial Director, Citec Polska Ltd, Katowice. July 1995 and December 1996.

Tadeusz Belerski, Journalist, *Rzeczpospolita*. July 1995.

Włodzimierz Bojarski, energy expert, Polish Academy of Sciences, Warszawa. July 1995.

Marian Chaber, Deputy Director, Mining Restructuring Programming Department, Environmental Protection Section, the State Coal Agency (PAWK), Katowice. July 1996.

Aleksandra Chodasiewicz, the Polish Ecological Club, Katowice. May 1996.

Michał Downarowicz, employee at the State Inspectorate for Environmental Protection in Poznań in the 1980s. Associated with the independent green movement in the 1980s. July 1995.

Alfred Dubicki, Secretary of the International Commission for the Protection of Odra against Pollution in Wrocław. January 1998.

Jan Dudzik, Deputy Director, Department for Environmental Protection, Kraków voivodship, 1996.

Jan Dulewski, Deputy Director, Environmental Protection and Deposit Strategy Department, the State Mining Authority (WUG), Katowice. July 1996.

Ryszard Fedorowski, Redactor for *Trybuna Górnicza*, a weekly magazine in Katowice devoted to the Polish hard coal sector. July 1996.

Bogusław Fiedor, Professor, Wrocław Academy of Economics. November 1995.

Krystyna Forowicz, Journalist, *Rzeczpospolita*, Warszawa. July 1995.

Radosław Gawlik, member of Freedom and Peace (NGO) in the 1980s, member of Parliament for the Freedom Union since 1989. Vice Chairman of the Committee for Environmental Protection, Natural Resources and Forestry in the parliament. November 1995 and January 1997.

Piotr Gliński, Sociologist, the Polish Academy of Sciences, Warszawa. November 1995.

Krzysztof Göhrlich, Vice Mayor, Kraków. May 1996.

Stanisław Gołąb, Director, Siersza Power Plant, Trzebinia, Katowice voivodship. January 1997.

Iwona Jacyna, environmental journalist, *Życie Warszawy*. July 1995.

Roman Janiczek, Polish Power Grid, Warszawa. January 1997.

Wojciech Jaworski, Director, Department for Air and Soil Protection, Ministry of Environmental Protection, Natural Resources and Forestry. July 1995.

Zbigniew Kamieński, Deputy Director, State Inspectorate for Environmental Protection (PIOŚ), Warszawa. November 1995 and May 1996.

Bronisław Kamiński, Director for the Environmental Protection Department in Kraków 1980-88. Minister of Environment in 1989-90. Presently Director for Proeko (environmental consultant company). July 1995 and June 1996.

Zbigniew Karaczun, researcher at the agricultural university in Warszawa, vice chairman of the Polish Ecological Club in Warszawa. December 1996.

Andrzej Kassenberg, member of the Polish Ecological Club. Director, Institute for Sustainable Development, Warszawa. July 1995.

Jan Kazior, Deputy Director, Ekosol (company specializing in desalination for the hard coal sector). Kraków. July 1996.

Andrzej Klasik, Professor, Katowice Academy of Economics. July 1996.

Eugenia Koblak, Deputy Director, Department of Water Management, Ministry of Environmental Protection, Natural Resources, and Forestry. July 1996.

Jan Komornicki, Director of the environmental protection department in Nowy Sącz voivodship in the 1980s. 1990s: Member of Parliament (PSL - the Peasant Party), Chairman of the Committee for Environmental Protection, Natural Resources and Forestry. December 1996.

Józef Kozioł, Minister of Environment in 1988-89. Presently: Director, Bank for Environment Protection, Warszawa. November 1995.

Stefan Kozłowski, Minister of Environment 1992. Chief Environmental Advisor to President Lech Wałęsa in 1993-95. Member of the Polish Academy of Sciences and the Polish Ecological Club. August and November 1995.

Krzysztof Krogulski, desalination expert, the State Coal Agency (PAWK), Katowice. May 1996.

Wacław Kulczyński, Deputy Minister of Environment 1987-89. Employee of the Bank for Environmental Protection, Warszawa. November 1995.

Jerzy Kwiatkowski, Department for International Cooperation, Ministry of Environmental Protection, Natural Resources, and Forestry (1988-93). Since 1993: Consultant for Proeko (environmental consulting) in Warszawa. November 1995.

Witold Maciejewski, member of the Polish Ecological Club in Swarzędz. August 1995.

Waldemar Michna, Minister of Environment in 1987-88. Member of Parliament (PSL - the Peasant Party) in the 1990s. November 1995.

Maciej Nowicki. Air protection expert at Warsaw Technical University in the 1980s. Deputy Minister of Environment 1989-90. Minister of Environment in 1991. Director of the Ecofund since 1992. November 1995.

Hanna Ogulewicz, Department of Geology and Mining at the National Fund for Environmental Protection and Water Management. June 1996.

Sebastian Pejm, Technical Director, Łaziska Power Plant, Łaziska Górne. January 1997.

Grzegorz Peszko, Researcher at the Cracow Academy of Economics. Member of the Green Federation. July 1995 and December 1996.

Tadeusz Pindor, economist, the Academy of Mining and Metallurgy (AGH) in Kraków. May 1996.

Danuta Plinta, expert, Voivodship Fund for Environmental Protection and Water Management in Katowice. January 1997.

Piotr Poborski, Deputy Director, Institute for the Ecology of Industrial Areas in Katowice, chairman of the Upper Silesian Branch of the Polish Ecological Club. May 1996.

Edmund Potok, chief electrician at the Łaziska smelter in Łaziska Górne, between 1955 and 1970. June 1998.

Leszek Preisner, economist, the Academy of Mining and Metallurgy (AGH) in Kraków. May 1996.

Zenon Raczkowski, Director, the Department of Environmental Protection at the Nadwiślańska Coal Company (NSW), Tychy. June 1996.

Wojciech Radecki, environmental law specialist, Environmental Law Group, Polish Academy of Sciences. December 1996.

Jan Rogóż, Technical Director, Łagisza Power Plant, Będzin. January 1997.

Jan Rzymełka, Vice Chairman of the Parliamentary Commission for Environmental Protection in 1990-93. Deputy chairman, the self-governing parliament of the Katowice Voivodship. July 1996.

Jerzy Śleszyński, Warsaw Ecological Economic Centre, Economics Department, University of Warszawa. July 1995 and July 1996.

Jerzy Sommer, Professor, Environmental Law Group, Polish Academy of Sciences, Wrocław. November 1995, December 1996 and September 1998.

Jerzy Swadowski, Deputy Director, Department for Mining, Power and Fuels, Ministry of Industry and Trade. July 1996.

Andrzej Szeja, the State Coal Agency, Katowice. May 1996.

Darek Szwed, member of the Green Federation, Kraków. May 1996 and January 1998.

Janusz Szyszko, Minister of Environment since 1997. January 1998.

Albert Tien, US Consultant to the Institute for the Ecology of Industrial Areas, Katowice. April 1997.

Stanisław Wajda, Legal Adviser, Ministry of Environment, Natural Resources and Forestry. January 1997.

Jerzy Wertz, Director, the Voivodship Department for Environmental Protection, Kraków. January 1997.

Janusz Wójcik, Environmental Director, Jaworzno III Power Plant, Jaworzno. January 1997.

Anna Zawiejska, Vice Director, the Voivodship Department for Ecological Policy, Katowice. December 1996.

Janusz Żurek, Institute for Environmental Protection, Warszawa. July 1995.

Tomasz Żylicz, Professor, Warsaw Ecological Economics Centre, Economics Department, University of Warszawa. Advisor to the Ministry of Environmental Protection, Natural Resources and Water Management. July 1995 and November 1995.

List of main questions asked in Poland

National policy and sectoral policies

What kind of continuity and what kind of change do you discern in Polish environmental policy in the period 1970-97?
What are the biggest successes and failures in Polish environmental policy (a) before 1989, (b) after 1989?
Are there any regions that have been/are in the forefront in environmental protection?
What were the main environmental policy documents between 1970 and 1997?
How and why did these documents come about?
What were the main causes of environmental degradation in socialist Poland?
What were the main environmental issues in (a) the 1970s, (b) 1980s, (c) 1990s?
When and why did environmental protection become a political issue?
Which kind of issues was excluded in the policy process before and after 1989? Why?
Who were the key participants in the environmental policy-making process between 1970 and 1997? Who were excluded and why?
With whom did/do you personally cooperate? Who did/do you consider your opponent?
In what way have representatives of various economic interests interacted with environmental policy-makers in the period 1970-97? (a) In the policy adoption phase, (b) In the implementation phase.
Have industry representatives blocked any environmental policy reforms?
What, in your opinion, should be the proper relationship between environment and economy?
What are your preferences with respect to policy instruments?
What are the main weaknesses and the strongest parts of present environmental policy?
How should environmental policy in Poland be improved?
What is your opinion about Polish membership of the EU from an environmental viewpoint?
What has been the impact of international environmental conventions and cooperation in the period 1970-96?

Sectoral policies

How serious is the salination/air pollution problem?
How should the problem be solved?
Who brought up the issue? Why was it brought up?
What have been the main changes in the strategy for addressing the salination/air pollution problem in the past decades?
What general political, economic, and social changes in Poland have had most impact on air pollution/salination policy?

Appendix 1: Environmental Protection Administration in Poland in 1985

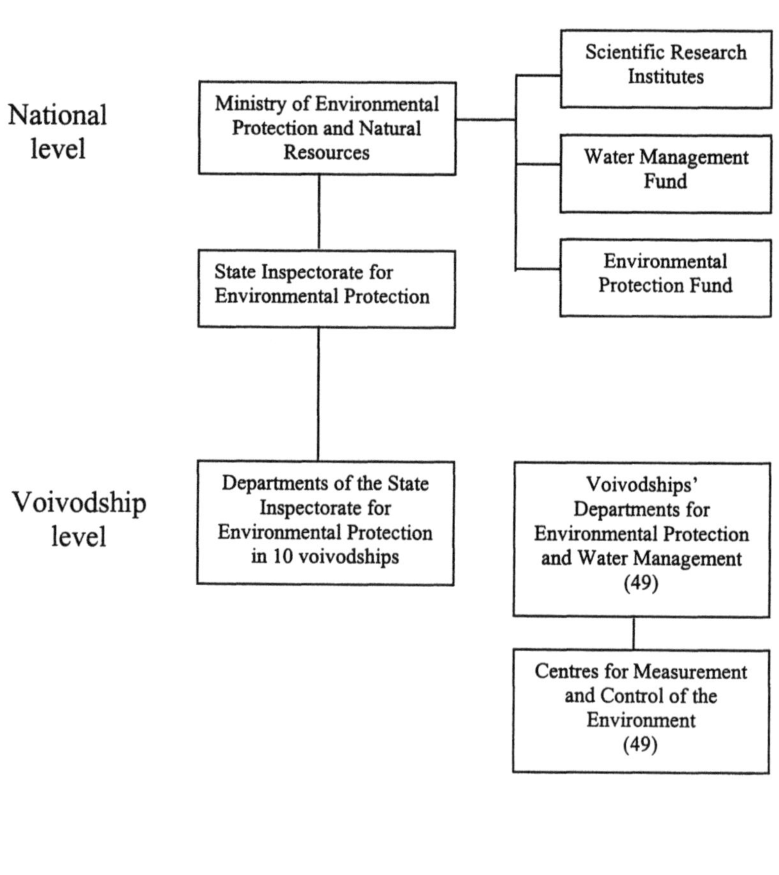

Local level

Environmental Protection Offices in the municipalities

Appendix 2: Environmental Protection Administration in Poland in 1995

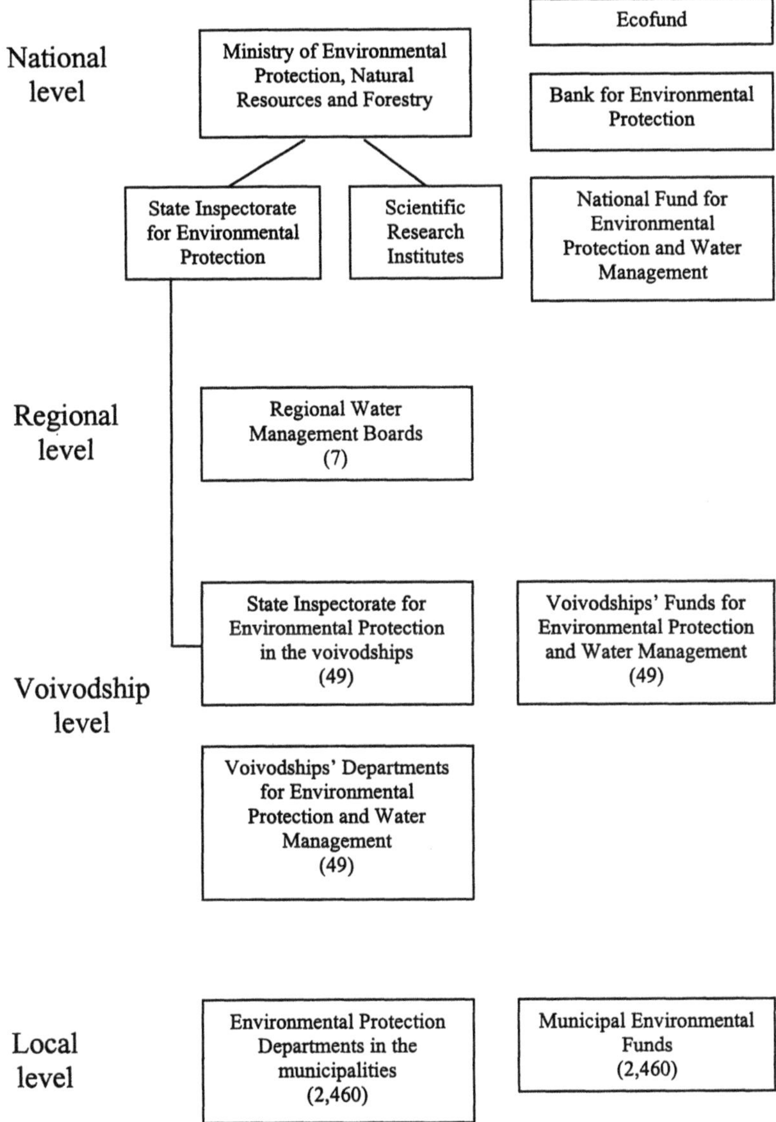

Appendix 3: Structure of the Ministry of Environmental Protection and Natural Resources in 1988

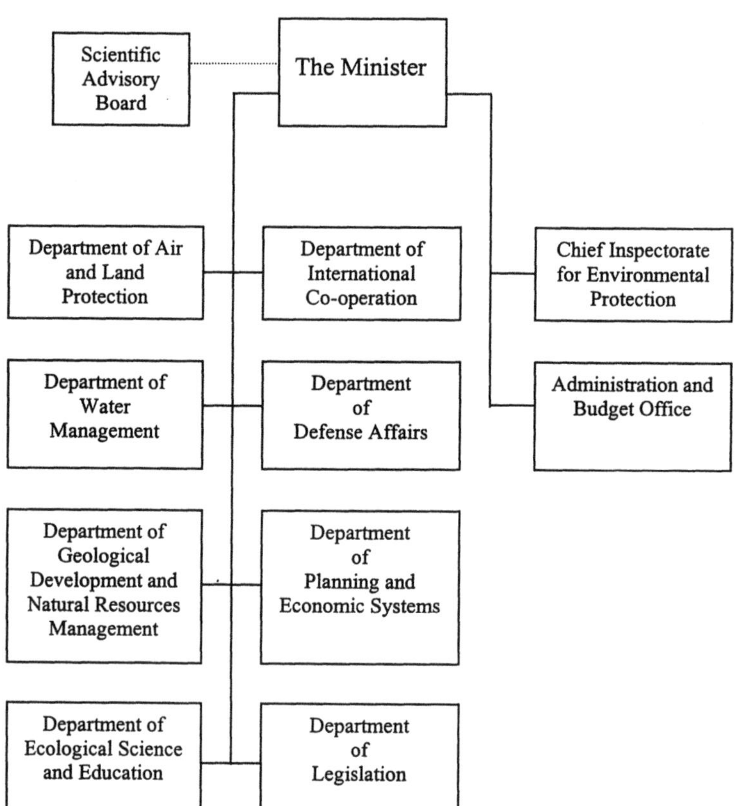

Source: MOŚZNiL, 1988c.

Appendix 4: Structure of the Ministry of Environmental Protection, Natural Resources and Forestry in 1995

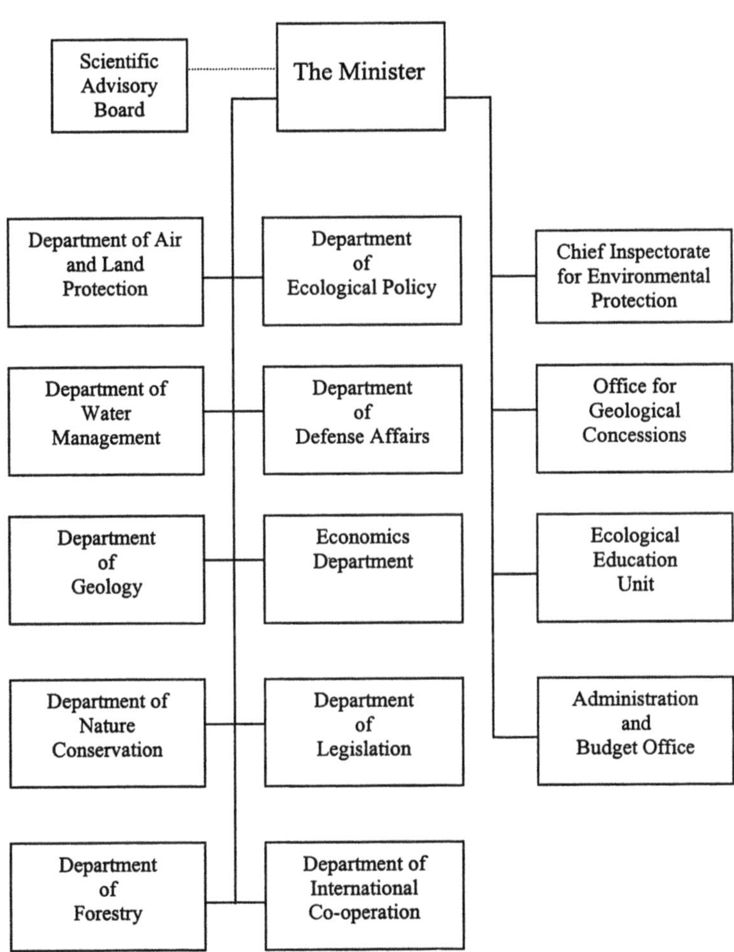

Source: OECD, 1995a.

Index

SAMENVATTING IN HET NEDERLANDS (Summary in Dutch)

Hoofdstuk 1

Het milieu was een belangrijk thema in de sociale en politieke transformatie die zich in de jaren tachtig voltrok in de landen van Midden- en Oost-Europa, in het bijzonder in Polen waar deze studie zich toe beperkt. De abominabele toestand van het milieu kwam onder meer tot uiting in bossterfte, radioactiviteit, vervuild (drink)water, bodemvervuiling, maar ook in (macro)economische verliezen en een aantasting van de volksgezondheid. Tegelijkertijd kan worden vastgesteld dat relatief grote gebieden niet werden geëxploiteerd en hierdoor hun bijzondere natuurwaarden hebben kunnen behouden.

Vanaf de jaren zeventig werd er in de meeste landen van Midden- en Oost-Europa een milieubeleid gevoerd dat, althans naar de letter, in sommige opzichten strenger was dan in de meeste West-Europese landen. Dat de beleidsinzet weinig effectief was, is wel in verband gebracht met het staatssocialistische politieke en economische systeem. De transformatie in deze landen had eveneens een politieke en een economische dimensie. Politiek gezien bracht de transformatie een democratische rechtstaat met een meerpartijenstelsel, economisch gezien de overgang naar een markteconomie en het loslaten van het stelsel van centrale planning.

De centrale probleemstelling van het onderzoek luidt:

> In hoeverre is de politieke en economische transformatie in Polen van invloed geweest op het milieubeleid in de jaren negentig?

Polen leent zich goed voor het onderzoeken van dit probleem. Er was al sprake van een relatief goed ontwikkeld milieubeleidsysteem voor de omslag in 1989, hetgeen een vergelijking tussen milieubeleid in de jaren tachtig en negentig vergemakkelijkt.

Hoofdstuk 2

De centrale probleemstelling wordt onderzocht aan de hand van vier aspecten van beleid, te weten (1) agendavorming en interactie tussen actoren, (2) de inhoud van het beleid (doelen, keuze van instrumenten), (3) implementatie en effecten en (4) de beleidstheorieën die aan het beleid ten grondslag liggen (*policy belief systems*). Het theoretisch kader dat zich goed leent voor een samenhangend onderzoek naar deze aspecten is het zogenaamde *advocacy coalition framework (ACF)* (Sabatier en Jenkins-Smith, 1993). Dit analysekader beschrijft de factoren die veranderingen in een beleid over een langere periode beïnvloeden. Het ACF is gebaseerd op vier premissen: Ten eerste, om beleidsveranderingen te kunnen waarnemen en verklaren zal beleid over een periode van zeker tien jaar bestudeerd moeten worden. Ten tweede, de analyse moet zich richten op een zogenaamd beleidssubsysteem, dat wil zeggen een geïnstitutionaliseerd beleidsveld gekenmerkt door een stabiel netwerk van actoren. Ten derde, de actoren, zowel gouvernementele als niet gouvernementele, vormen beleidscoalities (*advocacy coalitions*) teneinde het beleid in de door hun gewenste richting te beïnvloeden. Ten vierde, aan het beleidsgerichte handelen van coalities van actoren liggen expliciete en impliciete veronderstellingen ten grondslag die tezamen een beleidstheorie (*policy belief system*) vormen. In een beleidstheorie laten zich verschillende typen veronderstellingen onderscheiden.

Voor veranderingen in een beleid op de middellange termijn worden door het ACF diverse verklaringen gegeven: In de eerste plaats wordt gewezen op belangrijke gebeurtenissen buiten het beleidssubsysteem (bijvoorbeeld een regeringswisseling of een ramp). Ten tweede kunnen verklaringen gezocht worden in veranderingen die zich voltrekken in zogenaamde relatief stabiele parameters, in deze studie bijvoorbeeld het Poolse politieke en economische systeem. Daarnaast kan er sprake zijn van beleidsgericht leren binnen een beleidssubsysteem. Beleidsgericht leren kan plaatsvinden binnen een coalitie of in de interactie tussen coalities met conflicterende opvattingen en belangen.

Om de veranderingen in het Poolse milieubeleid en beleidsgericht leren sinds de jaren zeventig te beschrijven en te verklaren worden achtereenvolgens drie deelvragen beantwoord:

1. Welk milieubeleid is er in Polen gevoerd, zowel voor als na de transformatie in 1989 naar een politieke democratie en een markteconomie?
2. Welke veranderingen in het beleid waren incrementeel en welke radicaal?
3. Wat waren de drijvende krachten achter beleidsverandering en continuïteit?

Bij de beantwoording van deze vragen wordt steeds aandacht besteed aan de vier bovengenoemde aspecten; agendavorming, beleidsformulering, implementatie en beleidstheorieën.

Het ACF is niet eerder toegepast op een niet-pluralistisch beleidssubsysteem, zoals het Poolse milieubeleidsysteem in de jaren tachtig. Aangezien mag worden aangenomen dat de interactie tussen actoren in dit beleidssubsysteem een ander karakter had dan de interacties die het ACF op basis van de situatie in de VS veronderstelt, wordt in de studie ook gebruik gemaakt van elementen uit de politieke agendabouw-benadering (Cobb, Ross en Ross, 1976) en van de interactiemodellen die zijn beschreven door Ziegler (1987) in zijn studie naar het milieubeleid in de voormalige Sovjet-Unie. Bij de analyse van het Poolse milieubeleid wordt in het bijzonder gelet op pluralistische, staatscorporatistische (hiërarchisch) en sociaal-corporatistische (beperkte mogelijkheden tot participatie) interactiepatronen.

Hoofdstuk 3

In dit hoofdstuk staat de periode 1945-1980 centraal. Deze periode geeft met name inzicht in de relatief stabiele parameters en de ontwikkeling van het milieubeleidsysteem in Polen. Tot 1970 kunnen eigenlijk alleen natuurbescherming en waterbeheer als vormen van milieubeleid worden aangemerkt. De jaren zeventig laten de ontwikkeling zien van een milieubeleidsysteem. Dit blijkt uit de departementale verankering van het milieubeleid (1972), het vaststellen van een lange termijn milieuprogramma (1975) en de voorbereiding van een samenhangende milieuwetgeving. Toch vertoont de staat van het milieu in Polen in de jaren zeventig geen verbetering, integendeel. Milieubeleid was ondergeschikt aan verhoging van de industriële productie. Centraal in het beleidsvertoog stond de ideologische opvatting dat het socialisme per definitie beter is voor het milieu dan het kapitalisme. Deze opvatting werd gesteund door een flinke dosis technologisch optimisme. Het milieubeleid werd vorm gegeven door een kleine groep deskundigen los van het brede publiek en de industrie.

De belangrijkste impulsen voor de ontwikkeling van een milieubeleid in Polen kwamen van over de grenzen. Het Poolse milieubeleid werd in het bijzonder geïnspireerd door West-Europa en de Verenigde Staten. De belangrijkste externe gebeurtenis die de ontwikkeling van een milieubeleid in Polen versnelde, was de VN-milieuconferentie in Stockholm (1972). De binnenlandse milieuagenda werd in hoge mate bepaald door de agenda voor het buitenlandse beleid. Voor zover er in de jaren zeventig sprake was van beleidsgericht leren, kwam dit voornamelijk tot uiting in het imiteren van trends elders.

Hoofdstuk 4

In de jaren tachtig nam de aandacht voor het milieu snel toe en traden nieuwe actoren toe tot het milieubeleidsysteem. De eerste onafhankelijke milieuorganisatie, de Poolse Ecologische Club (*Polski Klub Ekologicny*, PKE) werd opgericht in 1980. In reactie op de kernramp in Chernobyl (1986) ontstonden nog andere onafhankelijke milieuorganisaties. Gedurende de jaren tachtig, zelfs tijdens de staat van beleg in 1982-83, was het milieuactivisten toegestaan tamelijk vrijuit hun denkbeelden in het openbaar te bepleiten. Het milieubeleid was een van de weinige beleidsterreinen waar participatie door de bevolking werd toegestaan en soms ook duidelijk effect sorteerde.

Het milieubeleidapparaat van de overheid werd uitgebreid. In 1983 werd het Staatstoezicht op de kwaliteit van het milieu ingesteld, in 1985 het Ministerie van Milieubescherming en Natuurlijke Hulpbronnen. Ook de media en wetenschappelijke organisaties deden van zich spreken. Met name de agendavorming van de jaren tachtig verschilde van die in de jaren zeventig: beleidsthema's werden niet meer uitsluitend van binnen het overheidsapparaat aangedragen maar kwamen nu ook van de oppositie. In de terminologie van de politieke agendabouwbenadering (Cobb, Ross en Ross, 1976), vond in de jaren tachtig een verschuiving plaats van een *inside initiative* naar een *outside initiative* model van agendavorming. Maar net als in de jaren zeventig was er in de jaren tachtig weinig directe interactie tussen actoren in het milieubeleidsysteem en actoren in andere (economische) beleidssystemen. Machtige economische actoren konden milieumaatregelen eenvoudigweg negeren, waarbij het *de facto* ontbreken van een rechtsstaat hen in de kaart speelde.

Opmerkelijk aan het Poolse milieubeleid in de jaren tachtig is de voorkeur voor economische beleidsinstrumenten. Beleidsnota's benadrukten de noodzaak van het integreren van milieu en economie. Tijdens de Ronde Tafel Conferentie van 1989, die de transformatie naar een politieke democratie en een marktstelsel markeerde, bleken zowel regering als oppositie het over veel punten eens voor zover dit het milieu betrof. De consensus kwam tot uiting in de omarming van een Poolse variant van het concept *duurzame ontwikkeling*.

Deze periode geeft een bescheiden verbetering te zien van prestaties op milieugebied. Maar de vervuiling door de industrie liep nog altijd uit de hand. Economische instrumenten misten hun werking door het ontbreken van een marktstelsel en handhaving was een farce door het ontbreken van adequate bevoegdheden.

Tegen het midden van de jaren tachtig beschikte Polen over een volgroeid milieubeleidsysteem. Het ministerie van Milieubescherming, de Poolse Ecologische Club en milieuwetenschappers in de Poolse Academie van Wetenschappen vormden het hart van de heersende milieubeleidcoalitie, in de studie aangeduid als de *Environmental Protection Advocacy Coalition* (EPAC). Kenmerkend voor de *beleidstheorie* van deze coalitie was dat de rampzalige toestand van het milieu in Polen niet werd gebagatelliseerd en dat de socialistische ideologie niet langer werd beschouwd als een remedie voor milieuproblemen. De EPAC bepleitte een verregaande integratie tussen milieubeleid en economisch beleid om de problemen het hoofd te bieden.

In deze periode was er dus geen sprake van verschillende coalities; beleidsgericht leren was het resultaat van de interactie tussen min of meer gelijkgestemde actoren van regering en oppositie die opereerden binnen het milieubeleidsysteem. Leren nam in deze periode een hoge vlucht, hetgeen tot uiting kwam in talrijke nieuwe beleidsinitiatieven. Cruciale externe gebeurtenissen waren de vorming van het onafhankelijke vakverbond *Solidariteit* (1980) en de kernramp in Chernobyl (1986).

Hoofdstuk 5

In 1989 voltrok zich de transformatie naar een democratie een en markteconomie. De Volksrepubliek Polen hield op te bestaan. Nieuwe milieugerelateerde beleidsthema's zijn privatisering, transport en consumptie. De regering lanceerde een radicaal economisch programma in 1990.

Waar het de inbreng van specifieke actoren betreft, zijn drie ontwikkelingen kenmerkend voor de jaren negentig. Ten eerste neemt de milieubeweging, ondanks een aanvankelijk sterke groei, een minder centrale positie in dan in de jaren tachtig. Ten tweede is er sprake van een toenemend aantal interacties tussen actoren uit het milieubeleidsysteem en het economische beleidssysteem. Ten derde wordt de inbreng van de wetenschap, mede als gevolg van de moeilijke financiële situatie, minder belangrijk. Agendavormingsprocessen vormen, evenals in de jaren tachtig, een mengeling van het *inside* en het *outside initiative* model.

Ondanks de economische teruggang werden in de vroege jaren negentig een aantal belangwekkende hervormingen doorgevoerd in het milieubeleid. Hiertoe behoren een sterke stijging van boetes en heffingen, het doorvoeren van maatregelen tegen de tachtig grootste (industriële) vervuilers en een zogenaamde *debt-for-environment swap* overeenkomst met enkele voor Polen belangrijke Westerse crediteuren, een overeenkomst met enkele landen die kwijtschelding van schuld inhoudt voor een bedrag dat in overleg besteed wordt aan een zinvol project op milieugebied. Het stelsel van financiering, handhaving, vergunningverlening en normstelling is ten opzichte van de jaren tachtig verscherpt.

De belangrijkste beleidsnota is het Nationaal Milieu Beleidsplan (1991). In dit document staat een pleidooi voor duurzame ontwikkeling centraal. Ook wordt gepleit voor invoering van verhandelbare emissierechten. Herhaalde pogingen om verhandelbaarheid te verankeren in de aangepaste milieu-kaderwet hebben tot dusverre schipbreuk geleden. Lessen uit het buitenland worden regelmatig gebruikt in de vormgeving van het milieubeleid.

In de jaren negentig is Polen beduidend beter gaan presteren op milieugebied, al is er nog altijd weinig gebeurd om milieueisen onder te brengen in sectoraal beleid. Een belangrijk motief voor Polen om het milieu te verbeteren, is gelegen in de wens om toe te treden tot de Europese Unie. Dit streven wordt door vrijwel alle politieke groeperingen gedeeld met uitzondering van het meer radicale deel van de milieubeweging.

Al in het begin van de jaren negentig valt het uiteenvallen waar te nemen van de beleidscoalitie EPAC die in de jaren tachtig de hoofdlijnen van het milieubeleid vorm gaf. Deze coalitie was nog sterk geënt op een gedeeld en niet altijd concreet besef van de ernst van de Poolse milieuproblematiek. Na de transformatie viel de EPAC uiteen in drie verschillende coalities. Deze worden in de studie aangeduid als de *Green Advocacy Coalition (GAC)*, de *Efficiency Advocacy Coalition (EAC)* en de *Mainstream Advocacy Coalition (MAC)*. De Groene Coalitie (GAC) benadrukt het belang van maatschappijverandering, in het bijzonder op het gebied van productie en consumptie. Aanhangers van deze coalitie hebben hun twijfels bij de zegeningen van de markteconomie. De Doelmatigheids Coalitie (EAC) maakt zich juist sterk voor een verregaande introductie van marktconforme instrumenten in het milieubeleid. De machtigste coalitie (MAC) opteert voor een middenweg, een versterking van bestaande economische instrumenten (boetes en heffingen) en flexibele handhaving, eventueel aangevuld met een bescheiden invoering van verhandelbare emissierechten. Beleidsgericht leren in de jaren negentig blijft beperkt tot de secundaire aspecten van de beleidstheorie. Het leren binnen coalities is pregnanter dan dat tussen coalities. Het verband tussen beleidsgericht leren en beleidsverandering is in de jaren negentig minder sterk dan in de jaren tachtig.

De bevindingen van de analyse van het nationale milieubeleid (hoofdstuk 3 t/m 5) worden in de hoofdstukken 6 en 7 nader tegen het licht gehouden aan de hand van twee gevalsstudies: luchtvervuiling door puntbronnen en de lozing van zout afvalwater door de mijnen in zuidwest Polen.

Hoofdstuk 6

Eisen met betrekking tot het terugdringen van luchtvervuiling door de industrie en de elektriciteitssector werden sinds het begin van de jaren tachtig geventileerd door de milieubeweging (PKE). In de tweede helft van de jaren tachtig kwam het vraagstuk hoog op de politieke agenda. Oorzaken waren de toegenomen wetenschappelijke bewijzen over de schadelijkheid van luchtvervuiling, acties vanuit de bevolking voor het sluiten van sterk vervuilende bedrijven en internationale druk. In de jaren negentig is de luchtkwaliteit, deels als gevolg van een betere implementatie van beleid, sterk verbeterd.

Het Poolse beleid om luchtverontreiniging tegen te gaan, steunde sinds 1980 op de volgende uitgangspunten:

- vergunningen verleend door de regiobesturen aan de vervuilers,
- normen voor luchtkwaliteit in de omgeving,
- regulerende heffingen voor wettelijk toegestane emissies en boetes voor overschrijdingen,
- heffingen en boetes worden ondergebracht in een fonds waaruit investeringen worden betaald om luchtverontreiniging tegen te gaan.

De drie beleidscoalities van na 1989 hebben alle deelgenomen aan de beleidsvorming rond luchtvervuiling in de jaren negentig. Dit geldt evenwel in mindere mate voor de Groene Coalitie (GAC) die zich met name roert in de discussie rond het toenemende transport. De Doelmatigheids Coalitie (EAC), bestaande uit enkele invloedrijke wetenschappers en leidinggevende figuren in de elektriciteitssector, maakte zich sterk voor de invoering van verhandelbare rechten. Ondanks enkele experimenten met dit instrument, liepen pogingen om dit instrument van een wettelijke basis te voorzien schipbreuk door de dominante positie van de Middenweg Coalitie (MAC). Deze coalitie bezet de sleutelposities op het ministerie van Milieu.

Evenals in het milieubeleid in het algemeen, vertoont het beleidsproces rond luchtverontreiniging in de jaren zeventig en tachtig corporatistische trekken. In de jaren negentig krijgt het een meer pluralistisch karakter. Hoewel er sprake is geweest van beleidsgericht leren, in de jaren negentig ook tussen de coalities MAC en EAC, heeft dit tot op heden niet geleid tot een herziening van de uitgangspunten over het beleid inzake het tegengaan van luchtverontreiniging.

Hoofdstuk 7

De mijnen in zuidwest Polen vervuilen de rivieren Wisła en Odra door het lozen van zout water. Verzilting schaadt het ecosysteem en de waterkwaliteit, hetgeen hoge kosten met zich brengt voor industrie en drinkwatervoorziening. In de jaren tachtig kwam dit probleem hoog op de politieke agenda. De mijnen werden verplicht heffingen te betalen ten behoeve van een nationaal fonds om ontziltingsmaatregelen te bekostigen. Dit instrument, dat in 1987 van kracht werd, heeft tot nu toe niet gewerkt, omdat de mijnen de hen opgelegde heffingen niet of niet geheel voldeden. Sancties van overheidswege bleven ook in de jaren negentig achterwege. De publieke aandacht voor het vraagstuk is sinds de transformatie eerder af- dan toegenomen. De mijnen verkeren in een niet al te beste financiële positie. Tegelijkertijd is kolenwinning het symbool van nationale zelfvoorziening op energiegebied, na de onttakeling van het Sovjetblok en bij gebrek aan import vanuit het westen voor de meeste beleidsmakers de enige optie. Mijnsluitingen zouden bovendien onaanvaardbare sociale consequenties hebben.

Het is opmerkelijk dat van pluralisme, kenmerkend voor het luchtverontreinigingsbeleid in de jaren negentig, geen sprake is waar het de verziltingproblematiek betreft. De Groene en Doelmatigheids coalities (GAC en EAC) hebben zich niet in het debat gemengd. Bij de aanpak van het probleem zijn uitsluitend actoren vanuit de MAC en de kolensector betrokken.

Het verziltingprobleem is ook een voorbeeld van beleid waarbij overlappende beleidssubsystemen - de economische en de mijnsector - betrokken zijn. De ongelijke machtsverhouding, in casu de dominantie van de mijnsector, verklaart in belangrijke mate de ineffectiviteit van het milieubeleid. De mijnsector is er in geslaagd de beleidsagenda te bepalen. Hier komt nog bij dat het verziltingsvraagstuk een nationaal vraagstuk is en er geen internationale overeenkomsten of EU-richtlijnen zijn die Polen tot een voortvarende aanpak kunnen aansporen.

Hoofdstuk 8

In dit hoofdstuk worden conclusies getrokken met betrekking tot veranderingen in (1) het beleidsproces (agendavorming en interactie tussen actoren), (2) de inhoud van beleid (doelen en instrumenten), (3) implementatie en (4) beleidstheorieën. Aan de hand hiervan wordt een antwoord geformuleerd op de centrale vraag in deze studie, de invloed van de transformatie op het milieubeleid in de jaren negentig.

Beleidsproces

Het Poolse milieubeleid van de afgelopen twintig jaar laat een mix zien van een pluralistisch, op incrementele aanpassing gericht beleidsproces en de twee corporatistische modellen die door Ziegler (1987) in de Sovjet-Unie waren waargenomen. Terwijl in de jaren zeventig een staatscorporatistisch model domineerde, gaven de jaren tachtig meer openheid te zien. Problemen werden van buiten het gevestigde beleidsnetwerk op de politieke agenda gebracht, er was sprake van een omslag naar een sociaal-corporatistisch model. Pluralistische interactiepatronen worden zichtbaar na de omwenteling van 1989, toen er ernst gemaakt werd met het vestigen van een rechtstaat. Hierdoor drong bij de industrie het besef door dat zij er belang bij had betrokken te worden bij het milieubeleid. Maar met betrekking tot het ontziltingsbeleid (hoofdstuk 7) worden nochtans geen pluralistische interactiepatronen waargenomen.

Consequentie voor de centrale vraag in de studie: de politieke en economische transformatie in Polen heeft bijgedragen tot een vedergaande openheid in het milieubeleidsproces, zij het niet over de gehele linie.

Beleidsinhoud: doelen en instrumenten

Het vigerende milieubeleid in Polen steunt op drie elementen: (1) een financieringssysteem gebaseerd op regulerende heffingen en boetes, waarbij opbrengsten worden geïnvesteerd in milieumaatregelen, (2) nationaal geldende milieukwaliteitseisen, (3) een vergunningenstelsel waarbij de provincies optreden als vergunningverlener. Dit stelsel lag in feite al vast in het Statuut ter bescherming en verbetering van de milieukwaliteit uit 1980. De jaren negentig geven wel een aanpassing te zien van de beleidsinstrumenten aan de nieuwe economische en politieke ordening. Het institutionele kader voor het Poolse milieubeleid is relatief stabiel gebleven over de bestudeerde periode.

Consequentie voor de centrale vraag in de studie: ten aanzien van het gevoerde beleid is er sprake van een opmerkelijke continuïteit tussen de jaren tachtig en negentig. Deze bevinding valt als volgt te verklaren. De Poolse transformatie in 1989 verliep betrekkelijk vreedzaam en zonder grote confrontaties. In brede kring heerste de overtuiging dat het milieubeleid dat sinds de jaren zeventig was ontwikkeld er op papier goed uitzag maar dat het ontbrak aan daadwerkelijke uitvoering. De milieubeweging had, vooral gedurende de jaren tachtig, actief aan de vorming van dit beleid bijgedragen.

Implementatie

Na de afschaffing van het socialistische systeem laat het milieubeleid een toenemende doeltreffendheid zien. De invoering van de rechtstaat leidde er toe dat ook de machtigste actoren, zoals overheidsorganen en industriële vervuilers, ondergeschikt werden aan het rechtssysteem. Het werd beter mogelijk om milieubeleid af te dwingen en te handhaven. De privatisering betekende dat bedrijven voor het eerst serieus aandacht moesten schenken aan kosten, waarbij zij niet voorbij konden gaan aan de milieuheffingen en boetes. De transformatie betekende ook het einde aan het relatieve isolement van Polen in de wereldeconomie. Deze verandering heeft ook gunstig uitgewerkt voor het Poolse milieu.

Consequentie voor de centrale vraag in de studie: de transformatie naar een politieke democratie en een markteconomie had zeer grote gevolgen op het terrein van de implementatie van milieubeleid.

De beleidstheorie

De beleidstheorieën van actoren hebben zowel in de jaren tachtig als negentig radicale veranderingen ondergaan. Zeer belangrijk was de omwenteling in het denken in het begin van de jaren tachtig (de opkomst van Solidariteit), waardoor het socialisme niet langer als onfeilbaar maar toenemend als de oorzaak van de slechte milieukwaliteit werd beschouwd. De opkomst van nieuwe beleidscoalities in de jaren negentig houdt rechtstreeks verband met de overgang naar een democratie en markteconomie.

Consequentie voor de centrale vraag in de studie: de transformatie droeg bij tot het ontstaan van nieuwe coalities met nieuwe beleidstheorieën die de weerslag vormen van de overgang naar een markteconomie.

Concluderend kan worden gesteld dat de veranderingen in het Poolse milieubeleid niet gelijkmatig zijn verlopen. Ten aanzien van agendavorming en proces, beleidsinhoud en beleidstheorieën hadden de meest radicale veranderingen zich al voor 1989 voltrokken. Na 1989 treden opnieuw radicale veranderingen op in het beleidsproces (zij het niet over de gehele linie) en in de beleidstheorieën resp. beleidscoalities. Veranderingen in milieubeleid waren radicaal in de jaren tachtig maar incrementeel in de jaren negentig. Alleen waar het de uitvoering van beleid betreft hebben de meest radicale veranderingen zich voorgedaan in de jaren negentig.

Verandering in beleid is vooral toe te schrijven aan externe gebeurtenissen, zoals de Milieuconferentie in Stockholm in 1972, de opkomst van het vakverbond Solidariteit (1980), de ramp met de kerncentrale in Chernobyl (1986) en de aanpassing van de Poolse wetgeving aan EU-richtlijnen in de jaren negentig. Van beleidsgericht leren is vooral sprake geweest in de jaren tachtig, toen het milieubeleidsysteem een coalitie kende waarin regeringsfunctionarissen en milieubeweging elkaar in toenemende mate vonden. Ook de jaren negentig geeft enkele voorbeelden van beleidsgericht leren te zien, in het bijzonder tussen beleidscoalities.

Tenslotte wordt geconcludeerd dat het ACF een bruikbaar kader is gebleken om het Poolse milieubeleid te bestuderen. De belangrijkste veronderstellingen die dit model articuleert, blijken niet minder op te gaan voor Polen dan voor andere beleidssystemen die met behulp van dit model zijn onderzocht. Wel zou de ACF verder kunnen worden versterkt door oog te hebben voor verschillende patronen van politieke interactie bij het structureren en agenderen van beleidsproblemen, zoals in deze studie is gebeurd.

www.ingramcontent.com/pod-product-compliance
Ingram Content Group UK Ltd.
Pitfield, Milton Keynes, MK11 3LW, UK
UKHW021828190726
13853UKWH00003B/1257

* 9 7 8 9 4 0 1 1 4 5 2 0 6 *